CATACLYSMS ON THE COLUMBIA

J Harlen Bretz, in his prime, typically wearing hard hat and field clothes, and filling his pipe with Edgeworth. Mt. Rushmore in the background. He titled this photo himself, "Five great men in one picture—four of them don't show." This must have been during the late '40s or early '50s.

CATACLYSMS ON THE COLUMBIA

A Layman's Guide
To The
FEATURES PRODUCED BY THE CATASTROPHIC
BRETZ FLOODS IN THE PACIFIC NORTHWEST

by

John Eliot Allen

Marjorie Burns

and

Sam C. Sargent

TIMBER PRESS
Portland, Oregon

DEDICATION

This book is dedicated to J Harlen Bretz,* who gave us a new concept and taught us ever so patiently that catastrophic floods may sometimes be a part of Mother Nature's repertoire.

S.C.S.

---

*Recipient in 1979 of the Penrose Medal, the highest award of the Geological Society of America.

ISBN 0-88192-067-3
Printed in Hong Kong

TIMBER PRESS
9999 SW Wilshire
Portland, Oregon 97225

CONTENTS

ACKNOWLEDGMENTS

This book is the culmination of my almost life-long interest in the Columbia River and the Columbia Plateau. I first studied geology at the University of Oregon and did my first field work in the Columbia River Gorge in 1931 under Dr. E. T. Hodge, who was involved at that time in the Bretz Flood controversy. Marjorie Burns is a professor of English literature at Portland State University. Her side interest in geology stems from backpacking and white water kayaking in the Northwest. Sam Sargent, whose expertise is woven throughout the book, worked for 5 years (1950–55) on the geology at and around the Dalles Dam. He was the first to propose a detailed stratigraphy of the Yakima Basalt lava flows and their interbedded sediments. He also assessed the quality and quantity of the sand and gravel deposits laid down by the flood waters for many miles up and down the river and picked those to be used as aggregate sources for the dam. In addition, Sargent located many of the erratics mapped in eastern Oregon and discovered the giant ripple ridges east of The Dalles. He has contributed 20 of the photographs used.

This book has been derived from more than 100 sources in the technical literature (see Bibliography), half of which discuss the origin and effects of the floods in Montana, Idaho and especially eastern Washington. In comparison, only 16 reports have been written on their effects in Oregon.

Early copies of the manuscript have been read critically by Ira Allison, Russ Bunker, Allan B. Griggs, Sally Allen McNall, Rhoda Bretz Reilly, Sam Sargent, Dave Taylor, and Cecile Woods.

The value of photographs in a book of this sort cannot be overestimated, and it is with great appreciation that we acknowledge the following sources (in order of number contributed):

Sam C. Sargent (20)
Oregon Department of Transportation (13)
Robert C. Carson (9)
Leonard Delano (8)
Ira S. Allison (5)
John Eliot Allen (5)
J. G. Rigby (2)
Harold Cramer Smith (sketch)
Soil Conservation Service (1)
Brubaker Aerial Surveys (1)

J. E. A. January 30, 1986

PREFACE

You Greeks are all children . . . you have no belief rooted in old tradition and no knowledge hoary with age. . . . You remember only one deluge, though there have been many. . . ."

—Egyptian priest to Solon, the Athenian lawgiver, in discussing the many catastrophes "by fire and water" that have repeatedly devastated humankind. From Plato, the Timaeus *(introducing the Atlantis legend).*

And the waters prevailed upon the earth
                                    GENESIS 7:24

INTRODUCTION

The idea of an inundation in the Pacific Northwest was first proposed by the pioneer geologist Thomas Condon, who postulated a "Willamette Sound" in 1871. J Harlan Bretz recognized the effects of catastrophic flooding on the Columbia Plateau of eastern Washington in 1919 and named it the "Spokane Flood" from a supposed but unnamed source north of Spokane. Later, when the great Montana lake which provided the water was identified, it was called the "Missoula flood." Evidence of multiple cataclysms* has developed over the years, so the plural term "floods" is now in general use. Most of this story has been published in technical journals (see Bibliography), but it has not heretofor been brought together in a story accessible to the general public.

It is now generally agreed that between 12,800 and 15,000 years ago more than 40 tremendous deluges of almost inconceivable force and dimensions swept across large parts of the Columbia River drainage. They were the greatest scientifically documented floods known to have occurred in North America. Nearly 16,000 square miles were inundated to depths of hundreds of feet. Swollen by the flood waters, the Columbia grew to contain ten times the flow of all the rivers in the world today and 60 times the flow of the Amazon River.

More than 50 cubic miles of soft silt, sediment and hard

*Webster: "1. A flood of water."

lava were carved out into a network of scabland channels, whose bare and eroded basalt surfaces and dry falls now typify large parts of the Columbia Plateau. The Willamette Valley was flooded as far south as Eugene, and where Portland lies today, the levels reached a height of 400 feet. Only the top 10 floors of the 40-story Interstate Bank Building would appear above the water!

Bretz's discovery and eventual vindication did not come easily. Considerable imaginative understanding was necessary to bring together the story of these floods and it was not a story that his fellow geologists wanted to hear. As a result, Bretz was faced with two challenges, first, the interpretation of the complicated geological events and, second, the arduous task of winning over his skeptical colleagues.

In this book we have used the name "Bretz Floods." This is not the first time an attempt has been made to honor Bretz's achievement by using his name. Back in 1952 there was a decision to call an unnamed coulee in the southern Hartline Basin "Bretz Coulee," but it was then discovered that the coulee already had a name, "Hudson Coulee" after two brothers who had attempted to start a farm in the area.

PART I, by Marjorie Burns, describes how geologists think and write and how they work as detectives solving the problems of a living, changing earth.

PART II, by Marjorie Burns, tells the Bretz saga—Bretz's early background, his first exposure to flood geology, his growing realization that the Columbia River drainage route had been greatly affected by colossal flooding, and, finally, Bretz's long battle to convince his colleagues that such flooding had occurred. This section also demonstrates geological methods of discovery by leading the reader through the process of Bretz's exploration and interpretation.

PART III, mostly by John Allen, is more scientifically detailed. It describes the landscape before and after the Bretz floods, tells how Lake Missoula originated, and summarizes the effects of the floods in the Grand Coulee and the Channeled Scablands of Eastern Washington, down the Columbia River Valley, into the Willamette Valley and on down to the Pacific Ocean.

PART IV, by John Allen, is a geological travelogue that extends from the source of the floods in central Montana to the Pacific ocean, describing the features en route. It enables travelers to find and observe them on their own along the northwest highways. The descriptions mention the U.S. and state highways, towns and cities, which are then listed for easy reference in the Index under "Highways" and "Cities."

APPENDIX A tabulates the areas of the ephemeral lakes produced by the floods.

APPENDIX B shows how Allen was able to calculate the height and relative volume of the greatest floods.

APPEXDIX C explores the magnitude of the energy generated by such flooding, and compares it with other catastrophic events such as earthquakes and volcanic eruptions.

The are two BIBLIOGRAPHIES, one for the lay reader and one for the technical reader.

Part I: Introduction

CHAPTER 1

THE LIVING, CHANGING EARTH

A word is not a crystal, transparent and unchanged, it is the skin of a living thought and may vary greatly in color and content according to the circumstances and the time in which it is used.

JUSTICE O. W. HOLMES

Why should anyone care for geology? There's no life in it, just rocks, things that stay put for the most part, and only a landslide, a sloughing of mud, a river swelling beyond its banks and rolling a rock here or there can make much of a difference. Well, admittedly, there are volcanoes, flows of lava, earthquakes. These things do shift the landscape around a bit, but for most of us such events seem more than a little improbable, like dragon sightings in the county next-door. Even for people in the Northwest, Mount St. Helens all too easily became something we saw for a while on T.V. So when we meet a rock, we don't usually think much of it. It's a leftover, a fragment, a chunk of something else, the rubble and scrap of bigger rocks.

We joke about rocks. A few years back pet rocks were in vogue: "easy on the furniture, non-shedding, guaranteed housebroken, may be fed on milk (it comes in quartz)" and so on. And this joking is understandable. Rocks as symbols of affection are amusing; and the prototype of the rock-lover, the rock hound, is—well, let's admit it—faintly comical.

But perhaps we're being unfair. Rock hounds are only the hangers-on, the groupies to the profession. What about the trained geologist? What about the degreed and practiced geological specialist? Surely some greater aura surrounds the professional, not perhaps an aura of romance and awe-inspiring wisdom, but some quality of expertise, something that commands our respect if not our emulation. And, yes, there is a special quality to the geologist. There's a lean, sun-parched leatheriness to most that sets them apart from other professionals. Even in the tweediness of an academic setting, the geologists are the ones with field boots disrupting the set of their cuffs.

Look closer at those boots. Therein lies (literally and figuratively) the understanding of the geologist. The geologist has roots in the earth; the geologist admits the clay in human beginnings. The geologist is a seeker, a wanderer, a questing philosopher; but the story of geology is as much a story of intrigue and polemic dispute as it is of dogged persistence and boot-paced exploration. Geology looks for the plot, the unfolding; it's the epic of sciences, the grand story of it all, the moving picture of the earth's history. Geology is not just a rock sitting still; it's process; it's change; it's transmutation. But we poor, quick-lived individuals, the drudges of the day, can't see that the tale is going on at this moment, that we're living within it. Like seasons to a May fly, geological changes are outside our time frame and virtually outside our comprehension.

"Well," you say, "That still doesn't make it much of a best-seller, does it? 'Process' isn't much of an audience-grabber. Where's the sex and violence?"

"Sex and violence"?—It's there, bigger and more screen-filling than ever you imagined. Doubters, sceptics! We're living, breathing, and walking around on a vast, changing, growing mass whose story, like the mating, birthing, and death of giants, makes our own small lives look puny indeed.

Our ancestors knew; they saw passion in the birth and death of rivers and mountains, oceans and worlds. They gave us the gods: gods-as-stars, and gods-as-planets, and gods of chaos, forces, and elements. "In the beginning" (said the Pelasgians in the third century B.C.), "Eurynome, the Goddess of All Things, rose naked from Chaos. . . . She danced toward the south, and the wind set in motion behind her seemed something new and apart. . . . Wheeling about, she caught hold of this north wind, rubbed it between her hands, and behold! the great serpent Ophion." She danced "wildly and more wildly, until Ophion, grown lustful, coiled about those divine limbs and was moved to couple with her." From this union came the sun, moon, and stars, the earth with all its complexities, and the distant planets, each to be ruled by a Titan and a Titaness.

Human beings, however, have never agreed upon any one version of the truth. Other say it was Uranus who fathered the Titans upon Mother Earth and that she, in anger, persuaded them to attack their father. Led by Cronus (who became "Chronus" or "time" for the Greeks) they did so; and coming upon Uranus as he lay sleeping, they castrated him with a sickle.

This sense of passion and violence in the creation and shaping of the heavens and earth is not merely a peculiarity of our ancestors. Even today geologists tend to write about their subject as though it were, somehow, made up of living, changing, and at times even willful beings, bestowed with emotions, purpose, and human features. The geologist's descriptive vocabulary is fraught with terms of violence, invasion, and resistance. Or it borrows from human physiology or from structures and buildings and common household equipment.

Glaciation "invades," "destroys," or "dominates." Floods, unable to "tolerate" normal water courses, "vigorously attack" or "overwhelm." Landscapes are "gashed," "broken," "sacrificed," or "overthrown," as though injury was deliberately inflicted. Topography can be "plucked," "scoured," and "scarred." Hills are "unyielding," and ridges "strongly expressed." Rivers are "fed" or "starved." Mountains, waterways, and other landforms have "heads," "mouths," "lips," and "skin"; "arms," "flanks," "feet," and "toes"; and those of you who have not checked the meaning of The Grand Tetons should do so. There are "floors," "benches," "basins," "entrances," "walls," "beds," "pillows," and "blankets." Even a gravel fill can be "intimately related" to a geological process, and all of it can be spoken of in terms of "birth," "maturity," and "decay."

What we get, then, if we pay attention to geological terminology, is a scrambling of human-related words and a sense (subliminal but effective) that the earth is in some ways alive. To feel this, to see the earth in this light, is to give ourselves the vantage point of gods (or geologists), to step beyond the limits of our years and sense the quickening rhythms of a world.

CHAPTER 2

DETECTING THE CLUES TO THE PAST

... every geological phenomenon is determined by an almost immeasurable number of variables, horrifying in their complexity and in the number of their interrelations. Every formula that uses a limited number of variables is therefore but an extreme simplification.

MARTIN GERARD RUTTEN, 1955

This is a story staged in two time periods. In more than one sense, it's a detective story, involving the reconstruction of an event that occurred over 12,000 years ago and "one man's effort to change the verdict" in our own time. There is no human crime involved, no plotted and executed murder. Nonetheless, the key events of this drama are based on catastrophes of such magnitude that understanding the mystery continues to be a prime concern for those who come upon the scene today.

It is not unusual to think of geology in terms of detective work. To a large extent geology is routinely a matter of investigation and testing of hypotheses. This in itself does not make geology different from most other sciences, but geology, like criminal investigation, deals with events that have already taken place and works to reconstruct the order and cause of these events, using methods more speculative than testable. Physicists, chemists, and biologists can rely heavily upon laboratory experiments; astronomers and geologists, for obvious reasons, cannot. Beyond the analysis of moon dust or other more earthly rocks, there is less of their subjects that will slip neatly into a petri dish.

Though, ideally, geological discovery is a continual play-off between speculation and proof, between the ivory tower and the field, ultimately the site itself must serve as the final check. This means getting out and clambering around on the earth (or the moon!) Like the detective, then, the geologist must sooner or later, turn from laboratory interpretation, (or from sweeping overviews and armchair speculation) to examine once more the "scene of the crime." A rock proven by lab tests to contain gold or oil does not necessarily mean that the area it came from is rich in these materials, any more than a trace of arsenic in a corpse is proof that the host at last night's dinner party ought to be arrested. It's a matter of puzzle work, of using all available techniques to rebuild the scenario. Think of the paleontologist confronted by a scattering of bones. From these incomplete bits and pieces and disconnected parts, an animal is to be reconstructed. There's no stomach, no flesh, no hair; and yet—from knowledge of how present-day animal species work—a handful of ancient remains, filled in with a generous supply of plaster, can lead to a skillfully conjectured and convincing specimen.

When geologists attempt to reconstruct the history of the earth, they too work this way, with random parts and puzzle pieces glued together by an understanding of geological process. It requires a sharp eye for unusual or easily overlooked clues and a sharp mind that can imaginatively but intelligently conceive of how the formation under study might have come about. (To this add good legs and sturdy boots.)

And there's more. Like any good detective, the geologist must not allow love of a particular pet theory to override other working hypotheses. The geologist must consider multiple possibilities and then work to disprove them, one by one. The simplest explanation that holds up under repeated attempts to discredit it is the theory favored. (In the trade, this is known as "Occam's razor" or "the principle of parsimony.")

Notice too that the term "hypothesis" is used more often than "fact." Though there are certain well-established facts known to the science, geology by its nature must remain full of mysteries. It's hard enough to make a criminal confess all the details and impossible to make a landform do so. Conceptions are therefore likely to be more a matter of wise conjecture than of mathematical certainty, and verdicts are mostly reached by consensus. In geological matters this nearly always means publication, so that other responsible scientists, operating much like a jury, can work towards some form of agreement. At times, however, dispute becomes hot and lengthy, and continual "retrial" begins to seem all too inevitable. It is then that courtroom-like hostility and Perry Mason theatrics creep in. Unpopular theories may be discredited without fair consideration; pertinent information may be withheld in order to maintain a preferred line of reasoning; and thinly disguised methods of personal attack infest the proceedings. It happened with plate tectonics, and it happened with the Bretz Floods.

Part II: The Bretz Saga

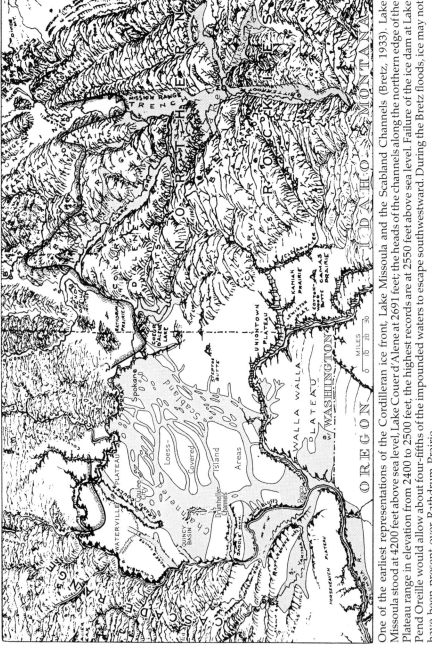

One of the earliest representations of the Cordilleran ice front, Lake Missoula and the Scabland Channels (Bretz, 1933). Lake Missoula stood at 4200 feet above sea level, Lake Couer d'Alene at 2691 feet; the heads of the channels along the northern edge of the Plateau range in elevation from 2400 to 2500 feet, the highest records are at 2550 feet above sea level. Failure of the ice dam at Lake Pend Oreille would allow about four-fifths of the impounded waters to escape southwestward. During the Bretz floods, ice may not have been present over Rathdrum Prairie.

CHAPTER 3

EARLY DAYS

We crack the rocks and make them ring,
And many a heavy pack we sling;
And run our lines and tie them in,
We measure strata thick and thin,
And Sunday work is never sin,
By thought and dint of hammering.

A. C. LAWSON, *"Mente et malleo,"* 1888

The geological detective tale that developed over the Spokane Flood Controversy is a tale of such sustained intrigue and epic proportions that its unfolding makes most fictional mystery plots seem trite in comparison. The two parts, both "crime" and "trial" (both the floods themselves and their interpretation over 12,000 years later) are impressive in their own right. They merge, however, into one story line through the work of J Harlen Bretz, the man who made their study a main focus of this life work and discovered the existence of the Spokane Floods. In our mystery novel analogy, it is he who plays the role of the dissenting but ingenious detective, the man who continues his pursuit of truth virtually unassisted and against all odds.

J Harlen Bretz spent over four decades defending his theories on the Spokane Floods before they were generally accepted by the scientific community. By then he was over eighty years old and, though limited in his ability to do field work, still mentally active. (He remained alert and productive until his death in 1981 at the age of 98.) His gift was the ability to take information and see it in a new light and then to check and recheck again and again to assure himself that his conclusions were correct. Bretz was above all a questioner and a thinker. And this, says his daughter, explains his reputation for rarely smiling. It was not from lack of warmth or humor, but simply that Bretz's mind was likely to be elsewhere.

The early details of J Harlen Bretz's life are relatively simple, but somewhere in his first years he clearly gained the insatiable curiosity and enduring stubbornness that later made him the man to recognize and document the Spokane Floods. He was born in Saranac, Michigan in 1882, the eldest of five children. "In my line, all had been farmers for a time at least," Bretz wrote about his boyhood. "As I grew older, my father Oliver thought that I might make a good stock farmer. I never agreed, went care-free through early boyhood and first encountered something far better, a small school library open only on Saturday afternoons."

From this came Bretz's first scientific fascination: astronomy. He built an observation platform around the central chimney of his "father's big two-storied red barn" and set himself to learning the names of stars, identifying constellations, and keeping observation notes. "But my first real discovery," Bretz claims, "was that I never could master the mathematics in astronomy." This shortcoming, the inability to work well with figures, was one that remained with Bretz throughout his life and one that Bretz jokingly liked to blame on an early head injury received from falling out of bed as a young child.

However much his father may have wished him to become a farmer, it is nonetheless clear that learning was encouraged in Bretz's family. Not only were his parents supportive of his astronomical hobby, but shortly after Bretz had completed public school, the family moved to the college town of Albion, Michigan so that he and the younger children following him could attend Albion College. Here, Bretz earned a degree in biology and met Fanny Challis, a classmate, whom he married in 1906.

Bretz's first training, it should be noted, was in biology rather than earth science. At this stage in his life there was no reason to believe his professional interest would change, but this in itself is not unusual. It is a curious and common characteristic about geologists that they frequently enter the profession by the side door, after having trained and worked in other areas first. Perhaps the explanation for this phenomenon lies in the nature of the discipline itself. Geology is one of those subjects that tends to appeal to us more as we grow older. For the young—wrapped up in their own immediate discoveries—it is easy to think of certain branches of learning as irrelevant, history, for example, and geology, subjects that seem removed from our own time. But for many of us, the older we grow, the more we begin to see ourselves as part of a continuing, growing, and changing system. We begin to take the broad view, to feel personally attached to history.

This "broad view," the sense of geological time working slow changes upon the earth, is perhaps what first began to fascinate Bretz. But whatever the cause, a geological passion began to overtake Bretz during the four years he spent in Seattle,

Washington as a high school biology teacher. The Seattle area, with its scars and signs of a Pleistocene ice age, is an excellent place to contract a case of geological infatuation. It was especially so in those days, since Bailey Willis, the crusty, outspoken father of Washington State's glacial geology, was present and promoting his interpretation of ice age effects on the local landscape.

It is no surprise, then, that Bretz—alert and inquisitive by nature—should began to gain an understanding of glacier mechanics. The smeared terrain left behind by the vanished ice ("foot tracks" of glaciers, you might say) intrigued him. He began to realize that much of the area south of Puget Sound is made up of giant, abandoned river channels, old water routes from the time when expanding glaciers nudged the drainage system southward. In this appealing and geologically rich part of the world, Bretz and his wife took to spending their free time traveling the area and making detailed notes on Pleistocene glaciation.

Now Bretz began to show his mettle. The weekend note taking became more than a hobby. His collection of field notes grew into scientific papers and were published in geological bulletins and journals. It was during this period too that Bretz mapped and named the Ice Age lakes around Tacoma and Olympia. This was no small feat! Remember that Bretz managed his geological exploration during the summer and in time spared from his job as a high school biology teacher!

Such an arrangement, however, was clearly less than ideal. Bretz needed advanced training in geology if he were to continue in the discipline. He therefore enrolled in the Department of Geology at the University of Chicago. His entrance into graduate school was partly supported by a fellowship, an indication of the outstanding professional reputation that Bretz had already earned. In 1913 he received his Ph.D., summa cum laude, with a doctoral thesis based again on his Puget Sound field notes. As an added triumph, this well-recognized piece of research was already in press even while Bretz was submitting it, in thesis form, to his supervising professors, a highly unusual accomplishment for a man just in the process of joining the profession!

In the making of this geologist, all steps seemed directed to preparing Bretz for his encounter with the Spokane Floods. He had cut his teeth on Pleistocene glaciation in an area of the Northwest uniquely influenced by the Ice Age. Puget Sound is not just a region that had been covered by mountain glaciers but an area filled with valleys and complex drainage channels repeatedly changed by the carving and crowding effects of the swelling and retreating ice. Bretz, through his Puget Sound investigation, had not only learned the movements and patterns of glaciers themselves but also how the outwash from a glacier normally behaves and how overflow channels work (where and how the melt-off and spillage would be most likely to run). It was excellent preparation for what lay ahead.

Dry Falls in Grand Coulee, with Fall Lake plunge-pool below the 300-foot high cliffs. The water pouring across this rim must have been at least 200 feet deep above the rim at flood maxima. (Allen photo).

CHAPTER 4

TOO MANY CLUES?

*In the inductive process, the more hypotheses the better. . . .
Contrary to this essential . . . the doctrine of
uniformitarianism leads to poverty where riches are
desired.*

HOWARD BIGELOW BAKER, 1938

Bretz's first real awareness of Eastern Washington's strange and intriguing geology came when he was still teaching high school in Seattle. He made it a habit to keep up with the most recent geological publications by visiting the Library and the Department of Geology at the University of Washington. During one of his frequent visits, he saw the recently published Quincy topographic map produced by the U.S. Geological Survey. The impressive dry falls (Potholes Cataract) showed up clearly at the western end of the basin. Bretz was already enough of a geologist to recognize the extraordinary oddness of such a formation. It seemed clear to him, as it had to others,* that vast amounts of water had once flowed through the Basin and out over the western end. But it was puzzling to find the upper lip of the cataract over 200 meters (600 feet) above the level of the present-day river. And the potholes themselves—large pit-like depressions lying beyond the falls—were of a size beyond anything known elsewhere: hundreds of feet across, much longer than they were wide, and deeper than a house is tall.

Now what did all this mean? Why, for example, should Bretz find the cataract so particularly surprising? First of all there was the matter of the falls being "dry," in itself puzzling, but the principal mystery lay in explaining the very existence of a waterfall in this unlikely spot. The fact is, geologists often refer to certain land formations, including lakes and waterfalls, as geological

---

*See Large (1922)

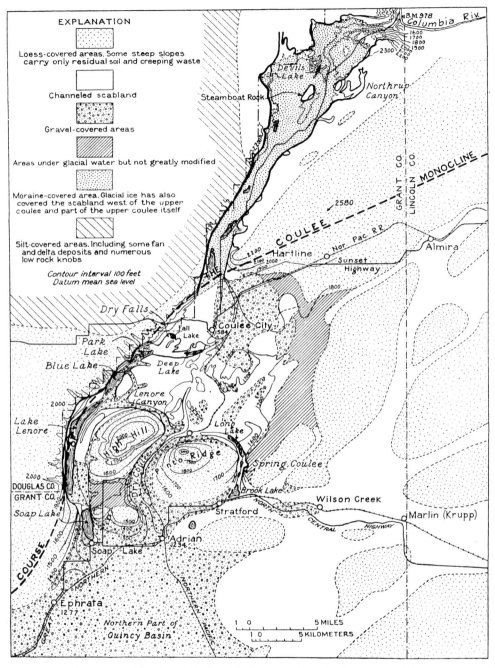

Geologic map of the Grand Coulee, Washington, drawn by Bretz in 1928. (In the explanation, the symbols for moraine-covered and silt-covered areas are transposed. The stippled pattern indicates silt, and the coarse cross-ruled pattern moraine.)

"accidents." They are temporary features on the earth, aberrations that normal geological processes will sooner or later eliminate.

Back in the late eighteenth century, James Hutton (1795) introduced certain geological theories that were taken up and further developed by John Playfair in a work that appeared in 1802 and was entitled *Illustrations of the Huttonian Theory of Earth.* The efforts of these two men established a main line of modern geological thinking (called Playfair's Law). They noted that rivers nearly always converge at the same level. Rarely does one pour into another from a higher altitude. What Hutton and Playfair observed about rivers and valleys is that geological changes come slowly—that with time all features tend to wear down to an evenness. Nature, through the occasional earthquake or volcanic upheaval, may create jagged forms that disrupt the smooth line of a river's flow, but they won't last. Erosion (working slowly, to be sure) will even things out, make gradients more gentle, drain off lakes, eradicate rapids, and pare down waterfalls.

This means, of course, that most of those natural features we consider vivid and picturesque are doomed. Waterfalls, rapids, and lakes are—geologically speaking—not long for this earth. In the process of aging, the strikingly contrasting and eye-catching features of a landscape tend to diminish, and a more uniform scene develops. Therefore, if we wish to find examples of geological "accidents," we had best look where the land is young: in mountains or foothills or on faulted terrain, places that have not yet been tamed by weather and time.

Consider, then, what Bretz felt when he recognized clear signs of what once must have been a gigantic waterfall north of the Quincy Basin. Here were broad channels and the remains of a cataract to out-rival Niagara but out on the surface of an otherwise flat or gently rolling prairie and away from any easily explained source of water! It was altogether a perplexing business.

There was, however, no doubt in Bretz's mind, or in anyone else's, that water—by one means or another—had indeed once flowed through the area. Indian stories explain that the Columbia River flowed over Dry Falls until Coyote, rejected in marriage, changed its course out of spite. Early white explorers and settlers, as well as the first scientists to visit the region, assumed that the Grand Coulee (as Rev. Samuel Parker wrote in 1835) "was indubitably the former channel of the river." Still, if this were so, why had the Columbia moved its path?

Other mysteries remained. There was also the matter of the potholes. But perhaps the term "pothole" itself needs some explanation before the size and mystery of those at Quincy Basin can be fully appreciated. For most of us, "pothole" conjures up the image of a poorly maintained, pockmarked road, and this isn't far wrong. But there is a more precise geological meaning to the word. What it amounts to is a pockmark in a *water*way, rather than a roadway.

Riverbed potholes are familiar enough to people who spend much time in and alongside steep, swift streams or rivers. Rafters, kayakers, and fishermen know them well. On "living" rivers (as opposed to the dry Quincy Basin) potholes reveal themselves mostly in the summer months when the spring run-off has drained away. Then they show best—to air breathing creatures—on rocky shelves along the banks; but underwater, along the path of the main current, they exist as well. They're smooth, bowl-like hollows that look as though they had been ground out of the rock surface. This, indeed, is what they are, and the grinding was done by the river itself, at times simply through the abrasion of its own swirling velocity, but more dramatically by using free pieces of rock, small river boulders, or even pebbles with a pestle-like motion in a natural mortar. In a groove or indentation— anywhere such a rock can be entrapped—swirling whorls of water grind the rocks against the sides of the hollow, eating a larger and larger hole.*

What we're speaking of, actually, are the effects of those circling, spin-off currents that appear within the main water mass, especially where an obstruction or side-stream causes a break in the regular flow of a fast moving, high gradient river. It's this action that provides the energy to grind potholes. Typically the pattern is complex, with swirls spinning within swirls until they spin themselves out, or as the geologist's doggerel goes:

> Big whorls have little whorls,
> Which feed on their velocity:
> Little whorls have smaller whorls
> And so unto viscosity.

Often, by the time we human visitors come upon examples of rock-ground potholes, the "pestles," the grinding rocks themselves, have been worn down and washed away or are mere grains of material lying at the bottom of the hollow, which has only to wait until the next wet season to capture new rocks. However, when the potholes still contain one or more good sized, well-rounded rocks, the perceptive river explorer has the satisfaction of

---

*The Dutch have a word for these: "Colk", the depression formed below a break in a dyke by the swirling waters.

Aerial view looking southeast from above Potholes Reservoir across Lind Coulee on the left, Soda Lake in the center, and the upper part of the Drumheller Channels, draining parts of the Quincy Basin on the right half of the picture (Delano photo).

Pothole lakes in the channels near Sprague, Washington (after Weis and Newman, 1974).

realizing how the potholes came to be; the two parts—the bowl and the worn-smooth rocks within it—are so clearly part of one system.

This, of course, is the simple story, the usual account of how potholes are formed in ordinary rivers, where they range from a few inches to six or seven feet across. These "everyday" potholes are intriguing in themselves, but they're insignificant compared to what one finds in Quincy Basin. There the potholes are gargantuan—giant, gaping hollows created by giant turbulence. Something more had clearly been at work here. And this was the point that troubled Bretz. If this "something more" had been simply an increased turbulence, where could such a turbulence have come from? And if some other effect had been at work, what could it have been?

The intriguing and markedly unusual features of the Quincy Basin topographic map were what first drew Bretz to focus on the geology of Eastern Washington. The second geological oddity that would later prove to be connected with the Spokane Flood presented itself to Bretz the year after he finished graduate school. He accepted a position as an assistant professor of geology at the University of Washington, an attractive option since a return to the Seattle area meant an opportunity to continue his study and exploration of the geological region he knew best, the region that had convinced him to become a geologist. For Bretz, firsthand contact with the terrain itself was always what mattered most, and appropriately enough, it was a field trip that now brought him in touch a second time with geological peculiarities created by the Spokane Flood.

The area Bretz visited on this influential field trip was the Columbia River Gorge, a stretch of steep river valley which cuts its way through the Cascade Mountains rather than creating a more likely (and less challenging) route alongside the uplifted range. This in itself makes the Gorge a bit unusual, but it was not the existence of the Gorge that most caught Bretz's attention. The U.S. Geological Survey map of Quincy Basin had mainly impressed him with the oddity of its dry falls; in the Columbia Gorge (which begins 120 miles southwest of the Potholes Cataract) what impressed and surprised Bretz were the number of misfit rocks. These misfits, called "erratics," are rocks located a considerable distance away from their area of origin.

The Spokane Flood erratics—as they would come to be known—are by no means limited to the Columbia Gorge. They occur within the full sweep of the flood's path, from Lake Pend Oreille to the sea and up Oregon's Willamette Valley as far south as Eugene. Altogether they're intriguing objects and easy enough to spot if one knows where to look. Some of the most eye-catching specimens (large, chalky-white boulders of granite) lie out on the farmland flats of Eastern Washington, near Ephrata and Soap Lake. The largest of these range (using rough comparisons) from the size of a pre-fab toolshed to that of a single car garage. There they sit—plunk—in the middle of fields, stolidly indifferent to the re-routing they cause in the otherwise straight and regular tractor furrows.

When geologists see large numbers of erratics, as Bretz did along the Columbia, and when some of these erratics are a thousand times larger than the river gravels transported by the Columbia today and are likely to be angular rather than smooth, they know something highly unusual has occurred. They know that these particular erratics have not been transported by the rolling, grinding, and polishing action of a river; these boulders have clearly avoided the usual breaking up and smoothing down that occurs to river-tumbled rocks.

It is at this point that the detective work begins. Facts must be added up and multiple possibilities considered even, at times, possibilities that may initially seem preposterous. What is it that could move rocks up to 20 feet in diameter and weighing up to 200 tons and scatter them randomly over an area as large as the Columbia and Willamette Valleys, an area thousands of square miles in size? Consider the possibilities. Could they have flown through the air? Absurd! But, still, perhaps they could have been hurtled through the air by a volcanic explosion. No, not rocks of this size; they were too distantly spread to have been spewed out of a volcano and were in the wrong locations and were composed of the wrong material as well. Could they be meteorites? Again the composition was wrong, and there were no craters to suggest meteoric impact. Could mud slides have moved them? No, not from the Rockies to the Columbia Basin, to the Willamette Valley and on to the Pacific! Could glaciers have pushed or carried them? Not here, not so far south. Could humans have moved them? Most unlikely! Even if there had been Indian tribes in the area at the time the rocks arrived in the Gorge, it is inconceivable that Pleistocene humans could have moved them such distances, and certainly there was no logical pattern to their distribution.

Bretz could think of only one feasible means of scattering such rocks randomly along this section of the Columbia River. They had floated in! Now, surely, of all possibilities, this one must seem the most improbable to the geologically uninitiated; but, in fact, this is precisely how the erratics found their way into the Columbia Gorge. They were moved by iceberg transport. Given a colder climate and higher waters than are now found in this part of the world, huge floating blocks of ice could have carried the rock material caught within their mass. When the icebergs melted, the rocks would drop out and settle by chance here and there, wherever water had been deep enough to support the drifting ice rafts.

In coming to this conclusion Bretz was following a line of reasoning that others had taken before. The idea of iceberg transport of rock and other material was not a concept that originated with him. It's a phenomenon familiar to geologists and the only one that logically explains the Columbia erratics (though nowhere else had erratics of this size been attributed to iceberg drift). It should also be pointed out that Bretz was not the first geologist to suggest that vast amounts of water had once inundated the Northwest. Forty years earlier in 1871, Thomas Condon, Oregon's pioneer geologist, had proposed a Pleistocene flooding of Oregon's Willamette Valley, but Bretz would carry these ideas further and in doing so would formulate theories that would bring him into extended and bitter conflict with the geological community.

Even at this early stage of speculation, there were certain obvious problems attached to the idea of high-water iceberg transport. The flood hypothesis, which was intended to explain the mystery of the erratics, introduced several new mysteries of its own! Where, for examples, could these hypothetical ice blocks have come from, and why was the water level at one time considerably higher than it is today? And how was it possible to account for the volume of water needed to float icebergs carrying erratics of such an immense size? It was this last matter that concerned Bretz most; his hypothesis demanded not just a moderate raising of the water level but an inexplicable and extraordinary flooding, a quantity of water that reason told him should not have been possible.

But these are questions that Bretz would not immediately begin to tackle. Professional difficulties had arisen, and Bretz left Seattle after only one year in the Northwest. The main reason for departing the University seems to have been his failure to find colleagues who shared his enthusiasm for field trips! Bretz was a man who took seriously the advice of the nineteenth century scientist, Agassiz: "Study nature, not books!" In his opinion, too many of his fellow professors preferred to practice geology from a desk or through a book rather than confronting it in the field. This was no small matter. Throughout his career Bretz would find himself contending with individuals who had not visited the region in question but who would nonetheless authoritatively oppose Bretz's position.

Fortunately, Bretz was invited to return to the University of Chicago (this time as a member of the faculty). And from here he soon established a program of summer field trips that took him back to the Northwest year after year, at first to the Columbia Gorge and later to Eastern Washington, that region he named the *"Channeled Scablands."* It was these summer field trips which gave Bretz his third encounter with the devastation of the Floods.

CHAPTER 5

SCARS WITH A DIFFERENCE

The rock scourings are the trails left by the invader. Their character should reveal the nature of the icy visitant as tracks reveal the track maker.

T. C. CHAMBERLAIN, *1888*

Bretz's own descriptions of the Scablands are dramatic:

"No one with an eye for land forms can cross eastern Washington in daylight without encountering and being impressed by the *'scabland.'* Like great scars marring the otherwise fair face of the plateau are these elongated tracts of bare, or nearly bare, black rock carved into mazes of buttes and canyons. Everyone on the plateau knows scabland. It interrupts the wheatlands, parceling them out into hill tracts less than forty acres to more than forty square miles in extent. One can neither reach them nor depart from them without crossing some part of the ramifying scabland. Aside from affording a scanty pasturage, scabland is almost without value. The popular name is an expressive metaphor. The scablands are wounds only partially healed—great wounds in the epidermis of soil with which Nature protects the underlying rock.

"With eyes only a few feet above the ground the observer today must travel back and forth repeatedly and must record his observations mentally, photographically, by sketch and by map before he can form anything approaching a complete picture. Yet long before the paper bearing these words has yellowed, the average observer, looking down from the air as he crosses the region, will see almost at a glance the picture here drawn by piecing together the ground-level observations of months of work. The region is unique: let the observer take the wings of morning to the uttermost parts of the earth: he will nowhere find its likeness." (1928)

By the early 1920's Bretz was well acquainted with the spell and complexities of the Scablands. He had only the summer months to devote to its exploration. During the academic year—in the months spent away from the Eastern Washington—he sorted through his growing collection of facts and puzzles and apparent impossibilities. It was no exaggeration that he spoke of creating an image of the Scablands "with eyes only a few feet above the ground." Bretz's earliest investigation was all by foot; he could not at first afford a car; but even later, when he acquired an early model, enclosed-body Dodge, much of the Scablands was nonetheless inaccessible to motor vehicles. Either way, by foot or car, Bretz and his party (typically composed of wife, son, daughter, collie dog, and a collection of students) were primarily limited to sighting across the broken and fragmented expanse of the Scablands. It was as though some infinitesimal mite were attempting to "read" by foot the chiseled message on a gravesite marker. Perspective was difficult to achieve and an overview all but impossible.

A friend and eminent geologist, Aaron Waters, describes what Bretz accomplished: "With Landsat photographs and detailed aerial photos it is easy for us now to see the nature of the scablands ... the giant ripples and other small-scale features 'prove the case.' But Bretz did this with the aid of only an agile and well-disciplined mind, and a pair of legs that were pushed over thousands of weary miles."

Bretz began by noting basic formations and studying the region's rock and soil make-up. It consists of an underlying basalt and a patchy covering of loess (a thick, wind-blown coating of ancient, Ice Age ground-up rock flour). And he saw, of course, scarring and channeling, but the full picture was missing, the overview that allows one to interpret the patterns and work towards an explanation. Any "whys" and "wherefroms" Bretz could offer were highly questionable and ill-defined. He had, however, his expanding store of information: bits and pieces, fragments and figures and facts. Like a jigsaw enthusiast who begins by sorting and organizing the scrambled pieces, Bretz was readying himself for the the time of assemblage.

In some ways it must have been a wearying and frustrating period, when so much of what Bretz saw didn't make sense. With land forms and erosional features all too often refusing to follow the rules that he had learned in his geological training, he was forced to approach matters in new and unproven ways. The vast

Scabland "canyon plexus" is a prime example. It was, in Bretz's opinion, "the most striking feature of the scablands," as well as the best indicator of the Scablands' origin. If these canyon mazes had been created by normal, slow, erosional processes, they would not show the braiding ("anastomosing," geologists say) patterns that they do. Normal rivers have a drainage system of small tributaries that flow into increasingly larger streams and rivers. Viewed from the air, this forms a tree-like, branching pattern (if we think of the full, consolidated river as the trunk and the multiple, small tributaries as the twigs). Here in the Scablands, however, canyons split and reunite and cross one another as though the water that had once flowed through them had been unable to make up its mind which path was best.

Clearly these were no ordinary drainage ways and could not be understood in normal terms. Bretz stresses this point with his repeated and emphasized statement that these are "channels, not valleys." Thus the significance of the name "Channeled Scabland." All of this, of course, had specific and somewhat troubling implications for the geologists of the times (and no less for Bretz than for his colleagues). They knew what is expected of a river valley and saw how much the Scablands deviated from the norm.

It works this way. Basically, river valleys are cut into the earth by slow erosional processes. A normal stream lowers its own bed over time, nestling deeper and deeper into its canyon or valley. In a sense, then, a river valley is a wedge of space eaten into the landscape, a wider and wider "V" spreading to the sky at a leisurely pace as the stream itself cuts through the land and as continuing rain and wind and gravity break down the existing edges and allow the loosened fragments to tumble down into the stream below.

It's a two part process, an interaction between slope erosion and stream transport. Underwater (and hidden from view) the channel is formed by the slow eating and deepening action of the current. At the same time, however, the steepened banks and slopes above the stream remain at an unstable slant. Bit by bit they fail and crumble and slide into the water. Now the stream's conveyor belt (or "transport system," as it's officially called) takes over and carries away the rubble, tumbling and washing along the current.

This is the usual picture. This is what a geologist expects to see in a normal river valley. But the Scabland canyons fail to comply. They fail to show the signs of continuous and long-lasting stream and slope interaction. In the Scablands there is too much a look of "gouging," as though some sudden and short-lived force had spent all its strength in one tremendous impulsion and cut neatly (in a single gesture) through whatever lay in its path.

To the trained eye, then, the Scablands look unfinished—as though events had stopped short. A "normal" river valley, as we said before, can be thought of as the empty space above a stream (measured from the water surface to the valley rim), but the Scabland channels must have been full-to-the-brim water conveyors, full-running trenches that had flowed over their tops in multiple locations making spillways and braidings between the larger channels and forming the Scabland's distinct and perplexing canyon network. There is no other way to account for these secondary, inter-connecting channels; the spillways show that water had clearly overtopped the rims of the primary canyons and spread out into secondary channels. What Bretz was faced with, then, was pure riverbed, from top to bottom. The waters that had flowed through these canyons had somehow failed to mine out a valley overhead. And this suggested to Bretz that whatever had happened had been sudden, vast, and short-lived.

"Distinct" and "perplexing" we have called this network, and indeed it is; and yet the whole matter can be looked at from another perspective. In a sense—if we change the scale—visible examples of this kind of channeling are not that unusual. Any sudden, heavy, flush of water starts by producing a channel. This is what happens before the water digs in deep enough to produce side slope erosion, and there are homely, modest examples of first stage channeling around us all the time. Those of us with city backgrounds have watched channels score their way through dusty, summer gutters when a neighbor up the street washes the car. And at the ocean such channels come and go with the tides wherever a small stream crosses the beach on its way to the sea. Each rising tide blurs the pattern; but when the ocean retreats, the stream once again cuts and braids new paths over and through the sand. It's a simple matter of water seeking its way over fresh terrain.

In fact, those with an experimental bent can create their own Channeled Scabland at home (and at virtually no cost). Lay a garden hose on a flat stretch of dirt and turn it on, a shade above a trickle. Channels form before your eyes, steep-walled mini-trenches weaving through the dampened dust or sand and splitting off and rejoining one another in an irregular pattern. These are not the marks of a stable, well-adjusted streamway but the effects of flood water seeking—hither and thither—the best and easiest path. No one route has yet claimed dominance, drawn most of the water to itself, and cut the deepest passage.

Now, if our hose is turned off and the day is sunny and dry, the newly formed micro Channeled Scabland will remain steep-banked and "anastomosed," at least until rains, winds, and time

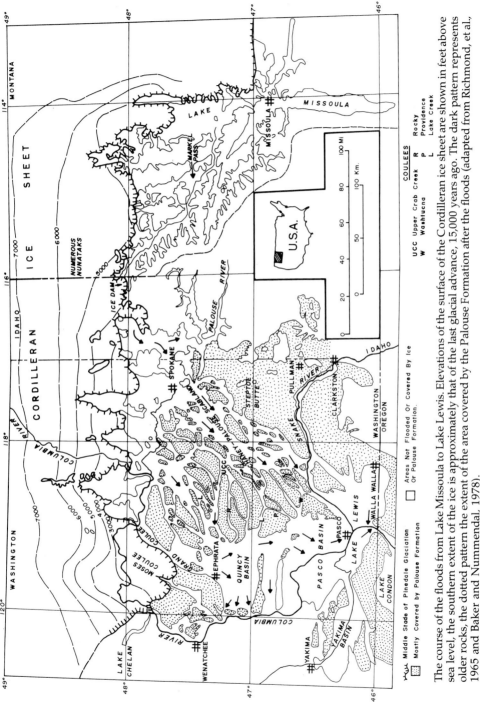

The course of the floods from Lake Missoula to Lake Lewis. Elevations of the surface of the Cordilleran ice sheet are shown in feet above sea level, the southern extent of the ice is approximately that of the last glacial advance, 15,000 years ago. The dark pattern represents older rocks, the dotted pattern the extent of the area covered by the Palouse Formation after the floods (adapted from Richmond, et al., 1965 and Baker and Nummendal, 1978).

collapse and obliterate the sharp vertical sides. With this picture in mind, an aerial photograph of Bretz's Channeled Scablands has a familiar and even commonplace look. It's simply another piece of water-scored ground. In scale, however, the difference is astounding, and Bretz, trekking through the vast coulees of Eastern Washington, lacked the god-like vantage point that the garden variety of channeling allows.

Channels, of course, were only one part of Bretz's growing collection of geological peculiarities. By the early 1920's he had begun to see any number of ways in which the Scabland and Columbia Valley geology were striking and extraordinary. The braided channels, as Bretz himself pointed out, were the most significant feature in his investigation, but there was also the matter of the ice-borne boulders—boulders of a material that Bretz had by now traced to an area many miles north of the Scablands. This rock material had somehow been carried southwest from its place of origin, through the Scablands and on down the Columbia Gorge to the Willamette Valley and the sea. The direction it had taken was clear, but the impetus that had moved the erratics was still unknown. It was an important but puzzling discovery, a clue and a mystery at once.

Granite erratic 3 × 5 feet in diameter, 3 miles south of Umapine in the Walla Walla Valley (Allison photo).

Then came the third feature in Bretz's collection of Scabland oddities. Like the Sherlock Holmes "Purloined Letter," this time it was something both obvious and easily overlooked. With matters of this sort, Bretz was at his best; he saw what others ignored and was quick to recognize the underlying significance! What Bretz began to notice were certain immense elevated hills or ridges *within* the Scablands channels. Unlike the surrounding terrain, these hills were singularly smooth and rounded in shape, but what was most surprising about them was their location. They lay high up within the channel complex but still below the upper rim of the canyon walls, an unusual place for such formations to exist. Certainly nothing in Bretz's schooling suggested their likelihood.

But Bretz was more than a book geologist. He had field experience—lots of it—and an independent mind. And he trusted his own perception. Remember that his first exposure to geology had been in the glacial outflow region of Puget Sound. There the Pleistocene rivers had been swollen to prodigious sizes and had transported, within their currents, larger than usual amounts of outwash debris. Piles of this debris (rock and sediment material) lie where they were long ago dropped by rivers that have diminished or disappeared. All rivers create such deposits, but those left behind by the Ice Age rivers south of Puget Sound are excessively large by the standards of rivers today.

East end of giant high-level gravel bar on the north bank of the Snake River west of Lewiston (Allen photo).

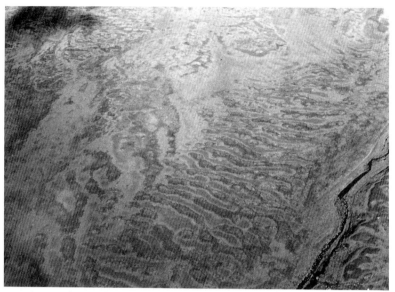

Aerial view of giant ripple ridges along the Palouse River (lower right) west of La Crosse (Carson photo).

Aerial view of giant ripple ridges in Lind Coulee near Providence Station (Carson photo).

Look at it this way: a man who deals with elephants is used to thinking big and a little more ready than the rest of us to accept something bigger still. Bretz's exposure to Puget Sound "elephants" gave him a lead. He began to realize that the Scablands' intra-canyon "hills" might just be larger-than-ever piles of river debris, "ultra-elephants," so to speak. They lay precisely in those places where—given an ancient flood—backwaters, eddies, and side currents would have allowed gravel deposits to form. If this were the case, those immense ridges and mounds tucked back here and there throughout the Scabland *coulees* (and, for that matter, down through the Columbia Gorge and far into the Willamette Valley) were nothing more than overgrown gravel bars.

When water rushes through a narrow canyon at high velocity, only the largest rocks and boulders stay put within the channel; the small debris is caught up within the current and flushes on through. But this changes when the water at last breaks free of narrow and confining chutes. When a river can stretch and spill itself comfortably over some broader expanse, with gaps and wings and side pockets along the banks, then backwaters and eddies are formed. These are the places behind promontories and obstructions where water can swirl around on itself at a lazier pace, take it easy, and drop its load of sand and grit. On ordinary rivers these are the spots we all like for swimming and picnicking (the quiet pool and the sandy beach).

There are other merits to river eddies. Since they're back behind what amounts to natural breakwaters or jetties, they create ideal resting spots for those who've had enough of fighting the heavier current. River runners use them as "parking spaces" (both along the bank and behind midstream boulders); and fish like them too, as shelters or "viewing stations" where they can lie in quieter water and wait for edibles to sweep past in the main current (like city folk in winter, tucked back out of the wind while they watch for the bus). And where there's an eddy, there's a deposit—small or large—of river-borne debris, rocks and sand piling up on the bottom and, above that, the slowly circling flotsam and jettsam of twigs and leaves or an all-too-likely styrofoam cup. Pull the plug, drain the water and what is left is a sediment mound.

Bretz's eddies were less obvious than those on "living" rivers. He had to envision where they would have occurred if vast amounts of water had indeed once flowed through the Scablands and on down to flood the Columbia. And the size of the gravel bars and their often surprising height above the canyon floor made them remarkably easy to overlook. There's a story that demonstrates the point:

Once, on a field trip through the Scablands, Bretz and three students were overtaken by night.

"Darkness," said Bretz, "found us well past the black sand dunes that dam back [Moses Lake], but open country still had enough night sky light to travel by. Then we encountered some ragged, rocky outcrops, around which Crab Creek became almost lost. Drumheller Ranch was somewhere among the craggy black basalt knobs, blades, pinnacles, rock basins and dry channels, but we missed it in the darkness. . . . Well, we stumbled, climbed, descended in the darkness for an hour or more, heartily sick of having attempted the traverse at night. I have since seen a topographic survey and aerial views of Drumheller Channels and unhesitantly give this area the palm for complexity of all flood-made topography.

"Fairly close to midnight, we saw a single light a mile or so ahead and found easier going. We were escaping that tangle of short channels, rock basins, basalt buttes; that is, getting across the backbone of the Frenchman Springs anticline. Of course, we aimed for the light. It proved to be a farm house. Reaching it, we saw through the unshaded windows a scantily clad woman fussing over two young children bedded down on the floor of bare boards. This wouldn't do, this Peeping Tom behavior. We shortly found the barn and a wagon loaded with newly threshed wheat. Sleeping on the wheat was the farmer. We awakened him, introduced ourselves and craved a refuge for the remainder of the night. We could sleep on the just-threshed wheat with him.

"The next morning he invited the four of us to breakfast. All that his daughter could provide was a carton of breakfast flakes. Glimpses of the interior of the house showed a table and a few straight chairs. Probably a bed or so that we didn't see. Otherwise, that house showed the direst rural poverty.

"We rode to Warden and a R.R. elevator. Enroute, the farmer told a bit of his misfortunes and stated firmly that, once he 'got a crop,' he was abandoning the farm. Years later, I saw his 'farm' again. The house and barn were gone, the field was gone, the site was now an active gravel pit. Yes, he had been trying to make a scabland gravel bar into a farm."

Gravel bars, erratics, and channeling: in themselves there is nothing unusual in these features. All three are manifestations that normal streams are quite capable of producing, in small to moderate sizes; but Bretz's erratics were gargantuan, his channels beyond anything known on earth, and his gravel bars were the size of substantial hills.

The conclusion to all this should seem obvious. Something sudden and prodigious had carved out those Scabland channels. Something had moved those erratics; something had piled up those gravel hills. And what could that something be, other than a massive (though still unexplained) volume of water, rushing over the Columbia Plateau and leaving nearly 3000 square miles of scarred and denuded land behind? As farfetched as a giant flood (in a relatively arid section of the world) may have sounded, there was nothing else that could explain the Scabland's extraordinary features. And this was certainly how Bretz found himself interpreting the Scabland geology. Yet, in his first scientific paper on the area (published in 1923), he went to some effort to avoid mentioning the possibility of a giant Pleistocene flood, and his reluctance was based on more than scientific caution.

CHAPTER 6

THE UNTHINKABLE HERESY

Catastrophism had virtually vanished from geologic thinking when Hutton's concept of "The Present is Key to the Past" was accepted and Uniformitarianism was born. Was not this debacle that had been deduced from the Channeled Scabland simply a return, a retreat to Catastrophism, to the dark ages of geology? It could not, it must not be tolerated.

J HARLEN BRETZ, in Baker, 1978

Bretz knew that the very idea of catastrophic flooding would threaten and anger the geological community. And here's why: among geologists in the 1920's, catastrophic explanations for geological events (other than volcanoes or earthquakes) were considered wrong minded to the point of heresy.

Probably this seems a little surprising, at least to those of us not familiar with the history of geology. How could one man's hypothesis (an hypothesis affecting only a limited region on earth) threaten an entire discipline? Isn't it the duty of science to be open-minded? Well, of course, this is what we might expect, if we assume the scientific community is more rational than other segments of society. But scientists, fortunately or unfortunately, are as human as the rest of us and as prone to irrationality, limiting habits of thinking, and (saddest of all) professional spite. Bretz's initial hesitation is fully understandable. There are not many of us eager to invite the scorn and hostility of our peers and colleagues.

Here we need to back up and look at what it was, exactly, that made other scientists consider Bretz's thinking radical. It requires a quick historical run-down on the development of geological thinking:

In the beginning was *Catastrophism*. And Catastrophism was not at odds with traditional religious beliefs. It was a theory of geological change that remained compatible with Judeo-Christian doctrine. Time was restricted to the Biblical six thousand years since *Genesis*, and within this short time span all the geological developments in the earth were accounted for. The result was a theory of earth history that saw all geological formations occurring through sudden and catastrophic events, either at the time of the original Creation or subsequently, through the acts of a Divine Will responding with reward or punishment to human behavior. We have, then, a God who is omnipresent—attentive and responsive—and who is seen as the force behind any shaping or reshaping of the earth. He is a God who can divide or close the Red Sea at will or call up a pillar of fire. He can compel the earth to "open her mouth" and "swallow up" rebellious Israelites or produce a forty-day Flood, the very flood which produced—it was believed—all the sedimentary rocks on earth. There were even those who maintained, as Thomas Burnet did, as late as 1681, that the very existence of geological features is evidence of human sin and imperfection, that mountains, valleys, and seas are nothing more than blemishes and disfigurations heaped upon an earth that was once as smooth and even and perfect as an egg.

By the eighteenth century, however, geological knowledge had expanded, and the job of rationalizing a growing list of events within a small and fixed supply of time became increasingly hopeless. In spite of Catastrophism's compliance with Biblical doctrine and its compatibility with human egocentricity, it couldn't last. It couldn't maintain its check on geological speculation in the face of developing evidence that showed, for example, that sedimentation occurs repeatedly or that uplift and erosion are continually and steadily changing the surface of the earth.

The first clear inroad against Catastrophism came in 1788 when James Hutton published his *Theory of the Earth*. It was Hutton who pointed out that sedimentary rocks had not been created by a single forty-day flood. They had instead been deposited layer upon layer in seas which had repeatedly covered the land. It was he as well who showed that mountains are both slowly uplifted and slowly worn away and that the lifting and the wearing away can occur simultaneously. According to Hutton, then, geological process is steady, slow, and repetitious.

All this would seem to place Hutton in strict opposition to Catastrophic thinking. The Catastrophists thought so; and yet, if we look at the matter in another way, Hutton's picture of the world is not all that different from what the early Catastrophists envisioned. A time-lapse film of geological change, as conceived by Hutton, would satisfy any visionary's need for the drama of a six-day Creation or the compelling sweep and terror of a world-rending Apocalypse. But there was more than this. The two doctrines also shared certain concepts of change. Both saw sedimentary rocks as the result of water-borne deposits, and both perceived mountains as having been uplifted, but Hutton's theories required time—vast amounts of it—and an acceptance of slow, regular, "uniform" processes. (Hence *Uniformitarianism,* the doctrine that grew to depose Catastrophism as the prime geological progenitor.)

This new way of considering time was what most disturbed the Catastrophists. Huttonian thinking asks us to stretch our concept of earth history far beyond what comes easily to humans. We're self-centered creatures, for the most part. We tend to look at the world through the stopped frames of our own short experience and are troubled by the idea of a time scale which invokes the infinite and sets us thinking of vast, alien epochs, of geological periods unacquainted with our human species and unaffected by our great human to-do. And, of course, in the minds of some individuals, such a perspective can only be seen as an attack upon traditional religion.

It was, therefore, no small task to bring Uniformitarianism out of scientific circles and convince the general public to see time in a new way or at least to reinterpret the boundaries of Biblical time. In spite of the fact that Uniformitarianism answered previously unanswerable questions and allowed scientists to explain geological changes through steady processes (through wearing away or sinking; accumulation, depositing, or uplift) resistance to the idea was strong and changes in thinking came slowly. But changes did occur. By the twentieth century the idea of geological time was at least a familiar concept, if not a universally accepted one. Geologists were not quite home free, but their discipline had gained considerable support, and their theories were listened to with increasing approval.

Consider, then, what Bretz was up against. The very word "Catastrophism" was heinous in the ears of geologists. To think in terms of massive, precipitous changes (beyond the occasional earthquake or volcano) was unacceptable, and the very idea of a sudden, colossal flood smacked too much of Biblical thinking, of a return to Noah, the ark, and the fifteen cubit depth (22.5 feet) of water which drowned the world (*Genesis* 7:20). It was a step backwards, a betrayal of all that geological science had fought to gain. It was heresy of the worst order.

CHAPTER 7

SOLVING THE JIG-SAW PUZZLE

Every great advance in science has issued from a new audacity of the imagination.

JOHN DEWEY, 1929

It should no longer surprise any of us that Bretz felt some hesitation in proposing his Spokane Flood. He did not, however, hesitate long. The evidence was too compelling to ignore and Bretz too spirited to play the coward's role. In his first 1923 paper, the closest Bretz came to mentioning a flood was to suggest that the Scabland's excessive channel erosion would seem to have required excessive amounts of water. Before the year was out, however, Bretz published a second Scabland paper, and here he took the leap.

In this second paper Bretz wrote with a new intensity. One by one he ran through his arguments, exhibiting them, pointing out the Scabland oddities, and explaining how "approved" geological concepts fail to account for the size, positioning and very existence of the Scabland's aberrant features. He pointed to the gravel deposits and stressed their extraordinary size and their often surprising locations. Some are more than three or four hundred feet above the present-day coulee floors. What, other than colossal flooding, could have placed them there? He emphasized again the importance of the erratic boulders (with their unmistakable signs of an ice-borne origin) and the significance of the Scabland's gigantic potholes, those "basins and pockets," that were clearly more than the usual, rock-ground potholes. Here (Bretz claimed) the force of the turbulence itself had been enough to "pluck" out chunks of the underlying *"columnar basalt"* (basalt that had formed column-like fracture lines as it cooled). And Bretz realized as well that many of these basins were in fact not actual potholes but deserted waterfall "plunge pools."

"Plunge pools?" Well, yes. That's the proper name for the spot where the force of a descending waterfall strikes bottom and carves out a basin underneath. But a *deserted* plunge pool is one whose waterfall has retreated upstream by "eating" at its own upper lip, by wearing away the rock material at the edge of its own drop. (It is this process that is slowly cutting Niagara back into Canadian territory and spurring Americans into applying drastic "repair" techniques to their own half of the falls!) Now, when a falls has eaten far enough upstream, it leaves behind plunge pools that are no longer directly beneath the descending water; and the Scabland canyon floors are cluttered with gigantic examples of just such deserted pools, elongated, empty Pleistocene plunge pools.*

To this expanded picture of erratics and massive gravel bars, potholes and retreating waterfalls, Bretz included one more item: a detailed, overall map of the Scabland's anastomosing channels, a patterning that in itself indicates flooding. And after laying all this out for the reader, Bretz ended his second 1923 paper with a brief and unequivocal declaration.

> "Fully 3,000 square miles of the Columbia plateau were swept by the glacial flood, and the loess and silt cover removed. More than 2,000 square miles of this area were left as bare, eroded, rock-cut channel floors, now the scablands, and nearly 1,000 square miles carry gravel deposits derived from the eroded basalt. It *was* a debacle which swept the Columbia Plateau."

But even then the research was not over. By 1925 Bretz had written two more papers on Scabland geology, adding still more facts and details to his argument. Among this new stockpile of evidence was the matter of *hanging valleys*. And what, you ask, is a "hanging valley"? Well, a hanging valley is, simply enough, a valley that has had its lower end cut off so that the water, flowing along, suddenly shoots out over a precipice and plunges down to a lower level, to the floor of a crosswise and deeper valley below or into a lake or sea. What we're talking about are waterfalls, though in Eastern Washington these hanging valleys are mostly dry (save for rainstorm or snow-melt drainage).

Think of driving along a familiar country road that has always before brought you smoothly down to join the main highway. Suddenly you find yourself dropping off the edge of the pavement and crashing down onto the crossroad. Someone or something has lowered the surface of the major route but failed to ease and adjust your side road to the new level. This is what happens to a stream when a fault or a glacier or a series of waves splits off or scours out or undermines the steady, even, sensible descent it has made for itself over time (for waterfalls, as we mentioned earlier, are temporary aberrations and the result of mere

---

*See page (113) for further discussion of potholes and plunge pools.

geological accident).

The Scabland coulees are full of such valleys, mostly hanging empty like gutter spouts with nothing to do. The high, splendid waterfalls of the Columbia River Gorge are themselves nothing more than hanging valleys, old side streams that have been snipped off short and left suspended and unsupported. Today they pour helplessly out into the air where once they made their way steadily down an established slope to join the Pleistocene Columbia River in a V-shaped valley a mile or so narrower than the present valley floor.

Braided channels, erratics, gravel piles, and hanging valleys; all of these could be accounted for by Bretz's hypothetical Spokane Flood, but coming to such a conclusion was not so simple as all that. In spite of his firm statement that "it *was* a debacle which swept the Columbia Plateau," Bretz was sorely aware of an inadequacy in his argument. He said it often enough himself. The concept of a colossal flood struck him as "so extraordinary" that whenever he concluded such a flood must have taken place, "in a few hours" doubt recurred. "Somewhere must lurk an unrecognized weakness. Where is it?" Bretz writes. But even as Bretz asked this question, he knew where the "weakness" lay. It lay "in the hydraulics of the concept." A flood, simply enough, requires water, and Bretz had no satisfactory way of explaining how so much water could have suddenly appeared on the Columbia Plateau, appeared and then disappeared, because, as Bretz soon realized, the water had drained away as quickly as it had come. The cleanly defined lines, the prow-shaped sculpturing of the gravel bars and the steepness of the coulee walls argue for an abrupt departure of the water, and this means the Scablands today have been left almost unchanged from what they were 12,000 years ago.

By 1925, after having published four papers on the Scablands, all that Bretz could suggest as justification for his sudden and voluminous deluge was the possibility of increased run-off from glacial melt, from water welling out along the edge of the Cordilleran Ice Sheet and draining away across the southwest tilt of the Columbia Plateau. (This is the ice sheet which covered most of the Canadian Rockies during the Pleistocene and which spread, for a time, far enough south into the northern United States to divert the Columbia River through what is now Grand Coulee.) But the run-off Bretz proposed wasn't enough. Nothing but prodigious amounts of water could have formed the Scabland's scarred and channeled features, and it was hard, indeed, to conceive of glacial melt achieving such a level of flooding.

Puzzles were added to puzzles. If the flood had resulted

from excessive glacial melt, how could such a sudden, intense, and temporary melting be justified? An explanation was required; and Bretz, for lack of anything better, offered two rather farfetched possibilities, improbable ones but the best he could contrive. Either there had been a sudden and extreme (but short-lived) softening of the harsh Ice Age climate or a gigantic *"jökulhlaup"* had occurred, an under-glacier surge of ice-melting heat caused by volcanic activity. Both possibilities would have greatly increased the glacial melt-off, but, as Bretz well knew, there was no evidence that such climatic or volcanic events had occurred. A sudden amelioration of the weather would have left its mark elsewhere and in other ways, and there were no indications that Pleistocene vulcanism had occurred in the area Bretz needed for his hypothetical jokÜllhlaup.

Again and again Bretz was caught between defending and doubting his own conclusions. "That the Spokane flood occurred is clear," he writes. "All other hypotheses meet fatal objections," and yet the "magnitude of the erosive changes wrought by these glacial streams is nothing short of amazing." Repeatedly Bretz refers to this sense of amazement. "During ten weeks' study of the region, each newly examined scabland tract reawakened a feeling of amazement that such huge streams could take origin from such small marginal tracts of an ice sheet." And the size of it all! Not even Niagara Falls and its gorge "approach the proportions of some of these scabland tracts and their canyons."

The discrepancy between what must have occurred and what could not be accounted for was unsettling to say the least. And, of course, Bretz, as were his colleagues, was a product of Uniformitarianism thinking. He took no pleasure in the idea of undermining a doctrine that had served the geological community so well for more than a hundred years. Again and again in his writing he expresses distress over the Scabland features and their apparent failure to coincide with geological possibility. Yet, again and again he concludes that the Spokane Flood must have existed. He complains of being "repeatedly driven" to a "position of doubt," only to return "by reconsideration of the field evidence" to the concept of an enormous flood. "It is the only adequate explana-tion of the phenomena. These remarkable records of running water on the Columbia Plateau and in the valleys of Snake and Columbia rivers cannot be interpreted in terms of ordinary river action and ordinary valley development. . . . Enormous volume, existing for a very short time, alone will account for their existence. The field evidence all the way from Spokane to Portland, a distance of 300 miles, knits together in a consistent whole to support this explanation."

CHAPTER 8

THE GREAT CONTROVERSY

And so these men of Indostan
Disputed loud and long,
Each in his own opinion
Exceeding stiff and strong,
Though each was partly in the right
And all were in the wrong!

JOHN GODFREY SAXE, 1849

The whole matter came to a head in 1927. Bretz, in fact, had been anticipating a general outcry and some form of reprobation since he first proposed his concept of the Spokane Flood. It was inevitable that sooner or later the geological community would rise up and attempt to defeat Bretz's "outrageous hypothesis." But the counter attack, when it came, was clothed in a somewhat unexpected form. It amounted to being cordially invited to his own trial. And the jury was rigged.

Here's how it came about. The Geological Society of Washington D.C. extended an invitation to Bretz, asking him to speak on "The Channeled Scabland and the Spokane Flood." On the surface, this seemed fair enough, but there was a catch. The invitation had not been extended in a spirit of scientific receptiveness or genuine intellectual curiosity. Put crudely, the esteemed Geological Society of Washington was "ganging up." They had collected unto themselves what one geologist, (Victor R. Baker) called "a veritable phalanx of doubters." They would allow Bretz to make a public presentation of his ridiculous and offensive hypothesis and would then expose it for what it was.

On the whole, there was something of a David and Goliath nature to the event, multiple Goliaths and one lone David (whose ammunition was not yet fully in hand). Bretz's main adversaries were among the best known, most respected geologists of the times. Most had been in the profession longer than Bretz, and most were members of the prestigious U.S. Geological Survey (men such as W. C. Alden, O. E. Meinzer, James Gilluly, E. T. McKnight, and G. R. Mansfield.)

To add to the awkwardness and effrontery of the situation, some of these men had dealt professionally with Bretz at other times and in other settings. In fact, two of the opposing geologists had previously been in positions somewhat subordinate to Bretz. One (James Gilluly) had been a student in Seattle's Franklin High School during the time Bretz was a teacher there. Another (O. E. Meinzer) had been examined by Bretz during his Ph.D. finals at the University of Chicago. Of all the men who confronted Bretz at this 1927 meeting, these two, Gilluly and Meinzer, were the most openly hostile (something for the psychologically minded to ponder).

The meeting began with Bretz running, once again, through the evidence that had led him reluctantly but conclusively to believe in his flood. He listed the effects he had listed before, first those within the Scabland canyons themselves: the "labyrinthine" patterning of the entire plexus; the rock basins along the canyon floors ("thousands of them"), some as long as eight miles, some as deep as 200 feet; and the "hundreds of extinct waterfalls." He spoke about erosion of the "once continuous loessal cover," erosion so extensive that in large areas 100 to 200 feet of soil had been removed, leaving only a few "small isolated loessial hills." He emphasized the importance of the "trenched divides," those distinct interconnecting spillways between the main canyons, and explained how they were proof that "water must have been 100 to 300 feet above preglacial valley bottoms"; and the "great mounded masses of little worn basaltic gravel," found in protected, downstream nooks throughout the Scablands and Columbia Gorge, were further proof of vast, catastrophic flooding. Of these, the most striking examples occur within the Snake and Tucannon Valleys, where the overpowering flood waters had rushed *upstream* to form a delta bar five miles long and 100 feet thick. Last of all, at the Wallula Gap (south of where the Snake and Columbia Rivers join) there are signs of deep ponding, signs that this narrow gateway had, for a time, held the waters back before they could continue 150 miles further and create still more of the vast, rough-edged basaltic gravel deposits. Nothing, Bretz concluded, but "a great flood of water abruptly issuing from the Spokane icesheet" could unite all of these extraordinary Scabland features into a "genetic system"; any other hypothesis would depend upon a series of extremely unlikely coincidences.

The case for the flood had been made. Now the rebuttals began. One by one Bretz's opponents stepped forward and presented their objections and disapproval. All of these men based their responses on a belief that "slow" processes are the only answer, and all showed a tendency to speak with a weightiness and formality that contrasted sharply with Bretz's vigorous, enthusiastic, and, at times, almost homey use of language.

In other ways the responses varied. Some were kinder than others. E. T. McKnight and W. C. Alden, for example, seemed genuinely considerate of Bretz's feelings, though their disapproval was nonetheless evident. There is something in McKnight's response that suggests Bretz might be welcomed back to the fold if he were to come to his senses. McKnight was unquestionably opposed to the idea of a Spokane Flood, yet the most critical term he used for Bretz's hypothesis was "inadequate."

Alden's response was mostly a plea for more research, for more time to be spent studying the region. Coming from Alden, this was no surprise. He had a longstanding reputation for caution (one geologist claimed "Cautious" was Alden's middle name), and it showed in his presentation. Terms of hesitation and uncertainty saturate his response: "it would seem," "perhaps," "not yet well enough understood." He admits to having read "Professor Bretz's papers on the subject with great interest" but adds that he was left feeling that "the true explanation of the phenomena" has not yet been found.

Others were more accusatory than McKnight and Alden, and some were belittling to the point of insult. Gilluly, for example, claimed Bretz's hypothesis was "wholly inadequate," "preposterous," and "incompetent" and maintained a tone of such condescension that Bretz, in turn, referred to it as "scolding."

But the most decisive and challenging response, as well as the most insulting, came from O. E. Meinzer, the father of modern hydrology. It was Meinzer's belief that a glacier-swollen Pleistocene Columbia could easily have produced Dry Falls and that the Scabland's puzzling, high terraces had been created, simply enough, by the river cutting its way down to lower levels. In his words:

> "The Columbia River is a very large stream, especially in its flood stages, and it was undoubtedly still larger in the Pleistocene epoch. Its erosive work in the Grand Coulee and Quincy Valley, impressive though it is, appears to me to be about what would be expected from a stream of such size when diverted from its valley and poured for a long time over a surface of considerable relief that was wholly unadjusted to it. The dry falls in the Grand Coulee resemble Niagara Falls and are evidently the product of normal stream work. The deep gorge of the coulee below the dry falls was apparently excavated by the same orderly

and long-continued process of head-end erosion as the gorge below Niagara Falls, and it could hardly have been produced in a short time by a flood of whatever magnitude. . . . Having seen only this part of the region in which I believe the existing features can be explained by assuming normal stream work of the ancient Columbia River, I am naturally loath to accept a theory of an abnormal flood for the scablands farther east. Before a theory that requires a seemingly impossible quantity of water is fully accepted, every effort should be made to account for the existing features without employing so violent an assumption."

This is a splendidly revealing passage. Look at the choice of language! Meinzer is "naturally loath" to accept Bretz's "violent assumption" of "a seemingly impossible quantity of water." Something remarkably close to overt hostility has crept into Meinzer's presentation (and is oozing back out between his passages of geological analysis). But what is also striking is the intensity of Meinzer's insistence upon a Uniformitarian time scale. He twice calls for the sanity of "normal stream work." He claims that the Pleistocene Columbia "poured for a long time" over the Scabland surface (an "orderly and long-continued process"), and he is convinced that the coulees "could hardly have been produced in a short time by a flood of whatever magnitude."

This strong Uniformitarian sentiment is, of course, exactly what one would expect from the geological community during the 1920's. Gilluly, too, looked for "a much longer process than granted by Bretz," and Mansfield insisted upon geological "persistence" and "long periods of years." Even Alden, in his cautious way, remarked that "the problem would be easier if less water was required and if longer time and repeated floods could be allotted to do the work."

But this unquestioning loyalty to the Uniformitarian party line was not the only similarity Bretz's opponents shared. Notice that Meinzer openly admits having seen only a limited section of the Scablands (Quincy Valley, Grand Coulee, and Moses Coulee). The fact was, none of the opposition seemed to have felt that firsthand exposure to the Scablands was necessary for a solid defeat of Bretz's hypothesis. Most of the men had only superficially studied the area at best, and then mostly secondhand, through reports and articles written by others. McKnight limited his arguments to the area immediately surrounding White Bluffs. Alden and Gilluly had not visited the Scablands at all; and Mansfield felt that it was enough to have studied basalts elsewhere since, he assumed, the Scabland basalts would inevitably conform to the normal pattern.

Let us pause here and return to the quotation which heads this chapter, the concluding stanza of the highly appropriate and much quoted poem, "The Blind Men and the Elephant."

> And so these men of Indostan
> Disputed loud and long
> Each in his own opinion
> Exceeding stiff and strong,
> Though each was partly in the right
> And all were in the wrong!

The relationship between these famous blind men (each with his separate and limited understanding of the elephant) and the Spokane Flood opponents should be obvious, though, in fact, the wise old men of Indostan were somewhat better off. They, at least, had all had *some* "field experience," while too many of the wise heads of geology at the 1927 debate were lacking even superficial acquaintance with the Scablands.

It's easy to imagine Bretz's frustration. He had admitted there are unexplained matters; he had admitted that error in interpretation is possible ("Even a bed-side practitioner may err, I understand."); but he found it hard to accept the unswayable criticism of those who believed "history can be diagnosed readily at long distance." A recurrence of the problem that had reportedly driven Bretz away from the University of Washington had come back to haunt him. Confronting Bretz were some of the greatest geological minds of the era, and each was holding on tightly to his own untested opinion, his own fixed and unalterable "truth," based, at best, on fragmentary research.

"I believe that my interpretation of the channeled scabland should stand or fall on the scabland phenomena themselves," says Bretz, and not just on sections of the scabland but on the whole picture, the "generic system" that accounts for all the features! "It is in the remarkable interrelationships of the channeled scabland *ensemble* that the conception of a Spokane flood finds support." "There are," Bretz adds, "many apparently possible alternative explanations," but any of these would require "exceptional combinations of factors," and none of them serve to explain "more than one or two" of the Scablands' many remarkable characteristics.

Again and again, on point after point, Bretz replied to his challengers, showing that their questions were the very questions he himself had asked and had already tested in the field. To Mr. Meinzer's suggestion that tilting of the land surface may account for some of the various channels, Bretz replied, "I have tried since to apply this in the field. But I cannot find any evidence to support it." To Mr. Mansfield he answered, "I have had no success in fitting the field evidence to the idea of shifting dischargeways across the scablands. I cannot get the glacial streams to cross at Palouse Canyon, Devils Canyon, etc., without a ponding farther down the Columbia." (Always there is a sense of immediacy with Bretz, a sense of his having been there, of his having thought through and tested each fact, each possibility, each hypothesis, against the next.) In exasperation he ended his response to Mr. Mansfield with the following pointed injunction: "I hope that Mr. Mansfield and others of the United States Geological Survey will be able some time to study the channeled scablands in detail."

If the opposition's intention had been to bring an errant Bretz back into the fold of reasonable geologists or if their intention had been to send him crawling away, silenced and drummed out of the corps, they failed, for Bretz's response to this kangaroo court debate was an increasing irritation with those who judge without experience and those who cling to belief systems without considering the evidence. No one's position appeared to have softened after the debate. Bretz's opponents seemed, if anything, more obstinate, more resistant than before. "Perhaps," said Bretz, a little wearily and towards the end of the meeting, "my attitude of dogmatic finality is proving contagious."

Think of these men of science working so hard to make sure that Bretz's concepts could not be true! Were they any different from the early Catastrophists who fought for what they literally believed was a God-given reality? Bretz's opponents saw only what they had resolved to see. They were prime examples of the geological quip, "I wouldn't have seen it, if I hadn't believed it." Like members of a faith whose existence is threatened when one small variant is thrown into the system, so these men seemed to feel the entire discipline of geology, and all it stood for, would be sent crashing if Bretz's reinterpretation of Catastrophism were given serious consideration.

CHAPTER 9

THE REVISIONISTS

The history of science is not so much in progressive accumulation of facts, as in the progressive clarification of problems.

R. G. COLLINGWOOD

So the Great Debate was over, and nothing was resolved. Bretz went back to the classroom and back to the field. The others returned to their various duties and their various chambers and nooks and headquarters, as firm as ever in their convictions. In the minds of those who had formally responded to Bretz at the 1927 meeting of the Geological Society of Washington, the matter remained a settled issue. The idea of a Spokane Flood was preposterous. It was too sudden, too massive, too quick-lived to have occurred. And, after all, where could such a volume of water have come from? It was sheer nonsense and an embarrassment to the profession.

This is the moment in a detective thriller or a Perry Mason drama when the "turn" occurs, the moment of utter despair and sorry defeat when new evidence pops up, things gel, or a timely confession is torn from a guilt-ridden suspect. This is the moment when the court (open-mouthed with astonishment) is compelled to change its verdict, to reconsider an opinion that moments before had seemed irrefutable.

And was it so for Bretz? Yes and no. Something in Bretz's favor did occur at the 1927 meeting—or so the story goes, but it had no effect on the Debate itself. Apparently a small stirring took place, a stirring that would take thirteen years to come fully to light; and it happened, not at the speakers' platform among the official participants themselves, but out in the audience. In the middle of the debate proceedings, Joseph Thomas Pardee, a geologist who was personally familiar with much of Bretz's terrain, reportedly turned to his friend, Kirk Bryan, and confided the following small but significant message, "I know where Bretz's flood came from."

Now here was a puzzle! If, in fact, Pardee knew the source of Bretz's water, why was he unwilling to make it known? The very nature of scientific advancement—so we are told—demands open-mindedness and a chance to build on one another's discoveries, and yet Pardee was to remain silent on the issue for more than a decade. The explanation was simple enough: Pardee had a career and a reputation to maintain. It was risky for any geologist to show a hint of support for Bretz's offensive hypothesis; and, to add to the problem, Pardee worked for W. C. Alden, the champion of scientific hesitancy.

Back in the early 1920's, just before Bretz began his own study of the Scablands, Alden (then head of Pleistocene geology for the U.S. Geological Survey) had sent Pardee to Eastern Washington to study the area's unusual geology. By 1922 Pardee had published the results of his study, claiming that the Scablands had been formed by glaciation of an admittedly exceptional nature. There is nothing surprising in this; others before him had looked to glaciation as an explanation for the Scablands and its channels. But Pardee's thinking didn't stop there. During his study of the region, he had begun to entertain another possibility, one that was not so comfortably in line with acceptable geological theorizing. He had begun to give serious consideration to a possibility that the Scabland's features had been formed from the drainage of an ancient, glacial lake named "Lake Missoula." No such lake exists today, but during the Pleistocene it spread out capaciously among the inter-connecting mountain valleys of Western Montana.

As early as 1910 Pardee had been studying and writing about Lake Missoula. His expertise in the area put him, more than any other geologist of the times, in a position to see the two regions (Western Montana and Eastern Washington) as part of one inter-connected geological story. But it was not a story whose time had come. The geological community was no more ready to hear of catastrophic flooding from Pardee than it was from Bretz.

Alden likely played a key role in the down playing of the idea. In a 1922 correspondence to David White, chief geologist of the U.S. Geological Survey, Alden referred to Pardee's Scabland research and spoke of "significant phenomena" Pardee had discovered; but, he added, the results "require caution in their interpretation," and "the conditions warn against premature publication."

It's not difficult to imagine what Pardee was up against.

Shorelines of Lake Missoula, just east of Missoula, Montana. The highest shoreline is at an elevation of 4200 feet above sea level, nearly 1000 feet above the present town (after Weis and Newman, 1974).

Alden was not only his superior but also a man of considerable influence within the Geological Survey and a fair judge of how such a flood hypothesis would be received. Silence had its advantages, and Pardee kept the peace. Nonetheless, judging from his whispered aside to Kirk Bryan during the Spokane Flood Debate, he did not go so far as to abandon the idea entirely.

It would be easy to dismiss the silence maintained about Pardee's hypothesis as merely a nasty example of scientific supression, but—as is often the case—it wasn't that simple. There seems to have been something genuinely troubling about the Lake Missoula explanation, something that cannot be offhandedly dismissed as simply the result of Alden's habitual caution. The truth is, by 1928, within a year after the Spokane Flood Debate, Bretz himself began to toy with the idea of a Lake Missoula drainage. But he too seemed to find the the idea less than fully assuring. In a paper given to the Geological Society of American in 1930, he went as far as to cite Lake Missoula as a possible source for his long-elusive water, but he did not promote the idea with much enthusiasm.

Here's why: Lake Missoula was a product of the Cordilleran Ice Sheet which, during its last advance, began to spread southward, branching into the northern tip of the Bitteroot mountain range and so plugging the normal drainage route of the Clark Fork River. The waters backed up; a lake was formed, rising to 4,200 feet above sea level and reaching a depth of at least 2,100 feet. Since (as Bretz explains) "the easternmost heads of the Scabland channels" are "400 feet lower than the lowest floor of Lake Missoula," if the ice dam at the northwestern end of the lake were to have failed, most of the lake water would have been discharged westward and then south over the Pleistocene prairie that became Washington's Channeled Scablands. So far so good. But—alas—geological "proof" was missing. There was, as Bretz himself readily admitted, "very little field evidence . . . of such escaping water between the lake basin and the scabland channels."

By 1932 Bretz was again saying "the cause of the flood is not yet known." He had, at this point, virtually finished his field research on the Channeled Scablands, and in his final studies he mostly ignored the Lake Missoula possibility. It was an intriguing hypothesis but too questionable to meet Bretz's standards.

Those rival geologists who accused Bretz of hasty conclusions might well have considered the extreme caution he demonstrated over Lake Missoula. Here at last was a likely source for his elusive flood water, but Bretz, in the face of inconclusive evidence, was unwilling to treat the idea as a settled fact. Instead he once again focused his attention on the Scablands themselves, refining and clarifying field evidence that argued for a massive flood—whatever the source. By better understanding the nature of the Scablands, Bretz wrote, we may yet be able to "attack the problem of its cause." And so, with this, the Lake Missoula solution was laid to rest for eight more years.

Between 1932 and 1940—the date when Lake Missoula again entered the picture—some small changes in attitude toward Bretz and his Spokane Flood began to occur. Those who had been the most adamantly opposed remained so, but here and there other geologists, who had become curious both by the reports and the controversy, began to study the Scablands on their own. This doesn't mean, of course, that they were willing to accept Bretz's hypothesis as it stood, but they were prepared to give the matter fresh consideration. This was precisely what Bretz wanted: experienced geologists who would look to the field evidence rather than assuming—sight unseen—a Uniformitarian position.

Throughout the thirties, more and more attention was given to the Scablands. The 1933 International Geological Congress, for example, included a field trip to the region. Bretz himself planned to lead the trip (he wrote a guidebook for the occasion), but other business made this impossible. In his place, Ira S. Allison, from Oregon State College, became the guide.

It was, of course, no ordinary, run-of-the-mill field trip. It couldn't be, given the Scablands. But even more important, the participants were not credulous souls, willing to trot at the heels of their guide. Several were internationally famous geologists, and these men especially (Allison recollects) insisted upon looking for their own evidence and applying their own reasoning to what they saw.

Expeditions such as this one helped to create an open season on Scabland speculation. Throughout the 30's, a rush for new hypotheses was on. Most, to be sure, were still re-interpretations that attempted to avoid Bretz's Catastrophism; nonetheless, among these new "see-for-myself" geologists were a number who began to suspect that Bretz may not have been entirely wrong; he had made some big mistakes, of course, carried an idea too far, but there were, after all, some reasons for his delusion.

These were the modifiers (or "revisionists," as Victor R. Baker calls them), individuals who attempted to re-work Bretz's Scablands hypothesis and render it into something geologically palatable. Among these revisionists were three men whose interpretations made a temporary but significant impact on Scabland thinking. They were: Ira S. Allison, who led the 1933 Scabland field trip; E. T. Hodge, Allison's colleague from Oregon State College; and Richard Foster Flint, a Yale University authority on ice-stagnation.

Of the three, Allison was the most congenial, the one who seemed most concerned with pleasing all parties. He was quick to give Bretz credit for his thorough and accurate descriptions of the Scablands, but what is even more striking is his willingness to accept the idea that extensive flooding had, in fact, taken place. He even went so far as to adopt Bretz's name for the flood. Nonetheless, Allison's "Spokane Flood" (1932) is not the same colossal deluge Bretz had offered the geological world. Allison could only go so far in accepting Bretz's hypothesis. What he could not accept were the duration, source, and magnitude of Bretz's Spokane Flood. He needed something less devastating, something less dramatic, a quieter, slower, more orderly flood, an ice-jam flood, for example:

Grant the possibility of a "blockade of ice in the Columbia River Gorge." This would cause the water to back up from the point of the ice-jam, growing slowly deeper and more extensive until it reached upstream into the Scablands themselves. There, over time (and diverted by further ice blockades into a "succession of routes" at "increasing altitudes") this ice-filled flood water would be able to carve and erode a rolling Pleistocene prairie into that area we today call the Channeled Scablands.

It's an ingenious conception, in its way. It explains, as Allison himself says, "the scablands, the gravel deposits, diversion channels, and divide crossings as the effects of a moderate flow of water, now here and now there, over an extended period of time. It thus removed the flood from the 'impossible' category."

But how did Allison account for his original Columbia Gorge ice blockage? Elementary. Allison proposed a landslide, a massive sloughing of earth and rock, mixed with a partial freezing of the river. Next question: how did such a landslide come about? Well, things get a little touchy at this point. "Admittedly," writes Allison, "such an icejam is difficult." But, he adds, "it appears far more feasible than the sudden release of an overwhelming, catastrophic flood." Most important, "it extends the time for scabland development." Such advantages, Allison felt, "more than offset the difficulties of the icejam theory."

After Allison came Hodge (1934), and like Allison, Hodge's hypothesis involved ice berg drift, plus a complex system of glacier encroachment. According to Hodge, it was the combined effect of water-borne icebergs and advancing and retreating glaciers that caused the Scabland's extensive erosion and the unusual patterning of drainage routes.

Throughout the 30's, in a series of presentations and seminars, Hodge vehemently defended this glacier/berg-drift hypothesis (and with equal vehemence attempted to discredit Bretz's interpretation). But Hodge, unfortunately, never managed a finished, published argument in defense of his alternate proposal. Field evidence failed to comply with his expectations, and this made it particularly difficult to counter Bretz, who persisted in basing his hypothesis on facts.

This matter of field evidence, in fact, is neatly illustrated by an anecdote involving Hodge and Bretz. Back during the time of Hodge's strongest contention, a scientific society, meeting in Pullman, Washington, invited Bretz to be their key speaker. The topic, as might be expected, was the Spokane Flood. When Hodge heard of this affront to good taste, he demanded the program include an opportunity to debate with Bretz. Permission was denied; but Hodge, undaunted, went on to create a subsidiary group and arranged to have it meet at the same time as the original society. Next, Hodge talked Bretz into participating in an informal debate, to take place the day after Bretz had given his formal address. Bretz agreed, and the debate took place.

It started off with Hodge presenting yet one more reiteration of his complex glacier/berg-drift hypothesis. Then came Bretz's turn. He stood up and invited the audience to accompany him on a short field trip into the nearby Scablands, where the erosion and prow-shaped hills of loess could argue for themselves. Along the way, he demonstrated, with a kitchen sieve, that above a certain level the loessal covering was composed of a homogeneous silt but below that point basalt and granite particles were mixed in, a legacy of passing flood waters. Once again Bretz had based his defense on the land itself—a fitting rebuttal to an opponent who argued more from wishful thinking than fact.

Flint (1938), the last of the three major revisionists, was not the easy mark Hodge had been. For one thing, Flint did his field work, and his style was more effective. *And* he was backed by Ivy League authority, that united (and Uniformitarian) Harvard and Yale front which had dismissed Bretz as coming from the University of Chicago, "that Western trade school." Altogether, Flint was an opponent to be taken seriously.

In his published "reconstruction" of Bretz's research, Flint maintains such an easy air of self-assurance and generous collegiality that it's hard not to feel this is an individual who knows what he's talking about. Even more than Allison, Flint exerts himself to give high praise for the "extensive information and excellent descriptions" that appears in Bretz's thirteen papers on the "Spokane Flood." It's an impressive display of cordiality toward one's rival, full of appreciation for "the earlier work of the pioneer" from which "the later investigator always benefits." But it's also a little excessive, as though there's more show than sincerity, as though the old man is being deposed and the younger can afford to be generous. (It may help to know that Flint earned his Ph.D. at the University of Chicago while Bretz was on the faculty.)

Like Hodge, Flint based his revisionist interpretation on glacier encroachment, but he felt there was also some form of extensive ponding involved, most likely the result of Columbia River blockage (as suggested by Allison). In simplified form it goes like this:

During the Pleistocene, melting glaciers along the northern tract of the Scablands created a runoff that "scoured" its way into the basalt and carried off the eroded material. At the same time, "rising ponded water in the Pasco basin" reduced the streams' gradients so that the "gravel, sand, and silt" were no longer flushed on down stream. Instead, this sedimentary material was deposited within the Palouse Valley, choking not only the main valley but tributary ones as well and damming side streams so that they too began ponding. As the drainage routes filled, streams began to "aggrade"; that is, they began to flow over higher and higher beds of their own making, running on top of material that they themselves had deposited. This fill—along with the now-ponded side streams—caused overflows to occur between valleys, creating the spillways that Bretz attributed to the sudden discharge of a single flood. When at last the temporary ponding drained away, streams could once again cut downward and did so in stages, thus producing Scabland "terraces."

This, in a nutshell, is Flint's account of the Scablands, an account based on "a picture of leisurely streams with normal discharge." It's a complicated hypothesis and a more challenging one than either Allison or Hodge had presented, but it came at a time when Bretz was ready to go on to other matters. He had done his Scabland research, presented his arguments; and now, for lack of an assured water source, he was willing to leave the question of a Spokane Flood to be settled by others. He was, understandably, a little tired of the whole business.

Then 1940 arrived, and the American Association for the Advancement of Science arranged to meet that year in Seattle, Washington. The geological section of the program—judging from the paper titles—was heavily represented by opponents of Bretz's hypothesis. In addition, Flint had been selected to lead a field trip through the Scablands. This was a field trip that Bretz chose to avoid, though Flint specifically made a point of inviting him to attend. There was no sense, Bretz felt, in submitting to the insult of futher confrontation. His field evidence and publications, he explained again, were all the defense he needed.

Hodge and Allison were two of the key speakers at this 1940 meeting. Hodge once again presented his berg-drift version of the Scablands; and Allison, who still believed flooding (but of a non-catastrophic sort) lay behind the Scablands, gave a paper (published in 1941) contrasting his own hypothesis with Flint's. There were other presentations but no surprises. It was basically repetition of material heard before, and all firmly Uniformitarian. But then came Pardee, the same Joseph Thomas Pardee, who had whispered to Kirk Bryan during the 1927 debate; and Pardee had a flood up his sleeve.

CHAPTER 10

"UNUSUAL CURRENTS"

There are geological arguments and methods of thought,
which though based on a combination of dimly perceived
facts, partly controlled guesses, personal intuitions, and all
manner of nebulous factors, not excluding a species of low
cunning, are yet in a geological matter decisive.

H. H. READ, *1952*

Pardee was the eighth speaker in the geology session, a position in the program when weariness begins to set in and the audience is likely to be less than fully attentive. For a man who had no desire to rock the geological boat anymore than necessary, the situation had a certain advantage. Pardee, of course, was aware that no one expected to hear open support for Bretz at a professional meeting, that no one wanted to hear such a message, and he intended to detonate his bomb as unobtrusively as possible.

The title of his paper, "Unusual Currents in Lake Missoula" (published in 1942), gave no real hint of what was to come; there was nothing in the idea of lake currents to startle an audience and nothing unexpected in his opening paragraph. He spoke first of the lake itself and its origin, the encroaching "lobe of the Cordilleran glacier" which plugged the drainage of the Clark Fork River. He then described certain basic, measurable facts about the lake. Its altitude had reached a maximum "of 4150 feet" and its depth had been "about 2000 feet higher than the floor of the valley just above the dam." This was tame enough, but then came the unexpected. "The ice dam is thought to have failed," said Pardee, in the middle of his list of facts, as though the event were of no particular consequence. He did add that this "permitted a sudden large outflow," but the statement is so lacking in fanfare it's likely that the audience, even then, failed to recognize the full implication of what he was saying.

As Pardee continued, however, something of the magnitude of the event must have become clear. The nature of

Lake Missoula's intra-valley layout created a number of narrow passes where the velocity of the out-rushing water, according to Pardee's calculations, "reached a maximum of 9.46 cubic miles per hour." (By contrast a notorious 1937 Mississippi River flood at Natchez achieved only ".05 cubic miles per hour.")

All this is impressive, but where was the proof? How could Pardee be sure that "a sudden great outrush of water" had in fact burst through the Clark Fork ice dam? Something must have given him the confidence to speak out. Some new element must have been added to his understanding of Lake Missoula. True enough; there was a new element but nothing that initially sounds all that remarkable. Pardee, to put it simply, had discovered ripple marks left on the floor sediments of Lake Missoula.

Ripple marks? We all know what those are, don't we? They're nothing but those small, wavy, configurations that show up along the shorelines of lakes, rivers, or oceans, wherever the water has retreated a bit. At the beach they're found in the shallow, sandy hollows that appear at low tide: lumpy, washboard-like formations, strange to walk on and pleasing in their smooth, patterned regularity. And what creates ripple marks? Currents— it's all the result of currents flowing over the bottom and warping the sediments into smooth, parallel, ridge-rows.

This is where the "unusual currents" in Pardee's title come in. Back in the valley basins, east of where the Lake Missoula ice dam had stood, Pardee had discovered ripple marks to end all ripple marks. And they could only have been formed by currents of an inconceivable size. In a sense, however, Pardee had not discovered anything "new" at all. These ripple marks had been there since the Pleistocene. The catch was that no one, before Pardee, had truly seen the Lake Missoula specimens. They had been walked upon, measured, and described (Flint referred to their "mamillary undulatory topography"), but they had not been recognized for what they were. Like Bretz's giant gravel bars, they had neatly disguised themselves as hills, rows of long, stretching, rounded hills, up to 50 feet high and spaced up to 500 feet apart.

In geology, viewpoint can be everything, and scale is easily deceiving. "Ripple" is not a word that brings to mind something large. It's a word we use for configurations that allow an overview, that show us the whole pattern in a single glance. For Pardee overview came with flight. From the air, symmetry and form reveal themselves, topography shrinks down to a meaningful size, and what felt like a succession of rather undistinguished hills to the pedestrian is suddenly unmasked as a series of extraordinary ridges that have, as Pardee says, "the form, structure, and arrangement of ordinary ripple marks but are so large that the term

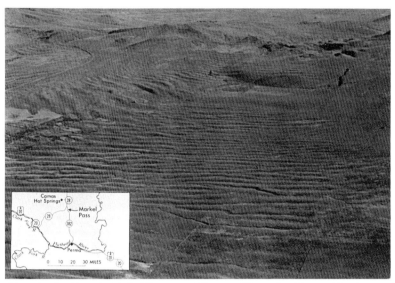

Giant ripple-ridges in Markel Pass, near Montana 28, south of Camas Hot Springs Montana. These mile-long gravel ridges cover an area of more than 6 square miles, they measure from 20 to 30 feet high and from 200 to 300 feet apart. The water pouring through this divide must have been 800 feet deep (Weis and Newman, 1974).

'ripple mark' seems inappropriate."

Once Pardee recognized the nature of his lake-basin "hills," he could begin to see other features in a new light. He began to realize that there were two types of gravel hills within the Lake Missoula boundaries: ripple mark gravel hills on the lake-basin floor and eddy deposits gravel hills that had been formed in the slack water behind promontories. He could now explain the extreme "denudation" that appears high up on the walls of Eddy Narrows (a ten-mile-long trench-like section of the lake basin); water rushing through this section had been at least 1000 feet deep and had moved at a rate of 78 feet per second!—or 58 miles per hour!

It all made sense. It all fell into place, once ripple marks had given Pardee the key to Lake Missoula's history. But of course, there was more to the matter than that. Pardee's ripple marks were as much a key to the Scablands as they were to Lake Missoula, and Pardee knew it. He was fully aware of what his ice-dam failure would mean to Bretz; and yet—strangely enough—Pardee made no mention of the Scablands being affected by his Lake Missoula drainage. He clearly chose not to connect his "sudden large outflow" with Bretz's flood, and his narrative stopped abruptly at the exit of the Clark Fork River, at the site of the ruptured ice-dam itself. Perhaps he wished to leave Bretz the honor of drawing the obvious conclusion. And perhaps, even at this point, he was a little unwilling to align himself openly with Bretz's unpopular hypothesis. But whatever Pardee's reasons for failing to trace his "great outrush" on its inevitable path to the prairies of Eastern Washington, the point was nonetheless made: the greater part of Lake Missoula's 500 cubic miles of water had indeed suddenly poured out across the panhandle of Idaho and inundated the Columbia Plateau in precisely that area which became the Channeled Scablands.

Was this, then, the moment of Bretz's vindication? Well, yes, to some extent. Vindication began here at any rate; but, in real life, victories are rarely of a piece and rarely come in a single moment. Pardee's 1940 revelation may have been the watershed in the change of thinking about Bretz's hypothesis, but official recognition came more slowly. The thirties had not seen the last of the die-hard revisionists. W. H. Hobbs, for example, (a well-known glacial expert from the University of Michigan) tried repeatedly throughout the forties to justify the Scablands through saner, Uniformitarian approaches. His solution was to bring a glacial lobe southward, down into the Scablands themselves where the erosive effects of the ice could grind out channels and push piles of gravel around. His first paper representing this idea was reviewed by those two adversaries, Bretz and Flint, and both promptly rejected it!

Even then Hobbs refused to quit. He kept on with a bullheaded, anti-Catastrophic persistence that did, after a fashion, pay off. He succeeded in getting his article into print, but first only in an abbreviated form in 1943 and later through a private printing in 1947.

In his "Foreword" to this article, Hobbs describes how he first became interested in the Scablands. During the winter of 1942–43, he heard Bretz give a presentation on the so-called "Spokane Flood" and was immediately struck, he tells us, by the obvious signs of glacial intrusion that showed up on the "lecturer's map." In a sentence or two, Hobbs offhandedly dismisses Bretz's interpretation and then moves on to speak at length of his own glacier-scour hypothesis.

The channels, according to Hobbs, are the result of ice gouging, and the gravel deposits are nothing more than "moraines" (the debris that is shoved along in front of an advancing glacier and then deserted when the the ice mass recedes). Against all geological logic, he speculates that the loess covering on the higher Scabland hills blew in while the glaciers were present.

All in all it's a classic example of single-track egocentricity. Hobbs is quick to tell us that his "Greenland observations" have given him "an advantage in interpreting the evidence within the Scabland region" and that the Geological Society of America was eager to provide him "with a grant of money which made possible a new study of the area." He goes on to explain that the results and conclusions of this field investigation, which occupied two seasons, "met with unusual enthusiastic general approval when they were presented to the Society in 1945 at its Pittsburg meeting. Following tumultuous applause in the crowded section the discussion was throughout approving."

Hobbs may have been the most egotistical of the remaining revisionists but he was not the last. The world has always been full of individuals who do not take kindly to being proven wrong. In the story of the Spokane Flood, Richard Foster Flint is the prime example of the species. In 1952 Bretz (now nearly seventy years old) once again undertook extensive field work in the Scablands, this time in the company of sympathetic fellow geologists. Afterwards, he sent a detailed field report to Flint, responding point by point to Flint's intricate "fill" hypothesis. In had been thirteen years since Pardee had sprung his Lake Missoula revelation upon the geological world, but even by then, Flint was in no mood to admit defeat. His response to the report consisted of a single sentence: "Scabland is proving to be pretty complicated."

The 1952 field expedition resulted in two significant publications, a lengthy 92 page article that appeared in 1956 and a shorter summary that followed three years later. More than anything, these two articles firmed up and reinforced all that Bretz had seen before, but it also added new ammunition to the flood argument. Gravel bars were studied in detail and excavated for content. Everything pointed to these massive river bars having been created at the same time as the channels and through the impact of extensive high current flooding. More giant ripple marks were discovered, this time within the Scablands themselves. These Scabland ripple marks, with their "recurring pattern of parallel ridges and swales," appear along the tops of the giant gravel bars, a clear indication that the bars themselves had been deeply submerged under a rapidly moving current.

But the most striking discovery that emerged from the 1952 expedition was the realization that the floods had been multiple, that it was no longer appropriate to speak of *the* Spokane Flood. Today, in fact, more than 40 floods are proposed, though Bretz, in the 50's, was convinced of only 7! What first suggested the concept of multiple floods, Bretz explains, is that certain "scabland buttes and basins" are clearly "more weathered" than others. These are the "older, higher channels," part of the anastomosing that occurred during the first flood, when the waters had not yet created a dominant path. Later, some of these channels were abandoned, "after main channels had become sufficiently enlarged to contain later floods." This process continued until, by the time of the last flood, the water "was confined to Grand Coulee alone."

Aerial view looking northeast to Palouse Falls (center). Note chasms extending north-south and east-west across picture, caused by flood scour along faulted zones in basalt. Remnants of loess occur only on highest mesas in upper right-hand corner of picture (Carson photo).

CHAPTER 11

VINDICATION

*Considering the nature and vehemence of the opposition to
this outrageous hypothesis, the eventual triumph of that
idea constitutes one of the most fascinating episodes in the
history of modern geomorphology.*

VICTOR R. BAKER, 1978

Altogether, the 1952 field expedition and the 1956 publication produced arguments that the geological community could no longer ignore. Even the most hard-bitten opponents of Catastrophism began to reconsider the wisdom of having judged Bretz without having visited the site. The result was that more and more geologists began to "throw away textbook treatments" (as University of Washington professor J. Hoover Mackin put it) and give the Scablands a fair and openminded goingover. And the Scablands *did* convince them—just as Bretz always said it would.

Among the new, growing list of converts was James Gilluly, one of the deans of North American geology. This is the same Gilluly who (without visiting the Scablands!) had so vehemently objected to Bretz's hypothesis back at the 1927 debate. Years later, when Gilluly finally took the opportunity to see the area for himself, he was startled to realize how little maps and measurements had done to suggest the actual size of things.

A. C. Waters was present and tells the story: "Gilluly was greatly surprised by the overall scale of the Scabland features: the breadth and depth of the water-torn coulees, the fills of coarse outwash gravel topped by giant ripple marks, the huge abandoned cataracts, dry waterfalls, and deep plunge-pool basins left by the Pleistocene floods. As he gazed upward to the 600 foot brink of Palouse Falls—over which the present diminutive Palouse River tumbles in a narrow 6 foot wide ribbon, making only a delicate tinkle, as it dimples the placid surface of the hugh plunge-pool below—Gilluly's expression changed from a look of astonishment to a broad grin. Clearly his mind had reconstructed the wild surge of a thrice-swollen Columbia River leaping from the brink with a roar of ten thousand Niagaras. Then he spoke loudy—perhaps competing with the noise of that vanished waterfall—"How could anyone have been so wrong?" His conversion—like Paul's on the road to Damascus—was sudden and complete.

There are other stories, equally gratifying. In 1965 the Seventh Congress of the International Association for Quaternary Research chartered a bus and arranged for an extensive tour of both the Lake Missoula region and the Channeled Scablands. Bretz—by this time nearly 83 years old and suffering from ill health—could not attend, but the tour nonetheless become for him an occasion of great personal satisfaction. Among the participants were several long-standing, outspoken critics of the flood hypothesis. But the Scablands did their stuff. As the trip progressed and more and more evidence in favor of catastrophic flooding was presented to the group, opinions began to change. "It was the last leg of the trip that did it," says Bretz, "and I knew it was all over when I received that long telegram of congratulations." It began with "Greetings and Salutations" and ended with one simple and gratifying sentence, "We are now all catastrophists."

Of all the conversions, this was the most pleasing to Bretz. But his victory was not yet complete. Not everyone who visited the Scablands was willing to shift over to the flood school of thought with the same grace and good humor as the International Quaternarians. There was still Richard Foster Flint. In the 1957 (second edition) of his well-known textbook *Glacial and Pleistocene Geology,* he ignored the Scablands entirely and referred only to a "temporary diversion of the Columbia River through the Grand Coulee." His own "fill" hypothesis had become something of an embarrassment since its sound defeat the year before and was noticeably missing from his revised text. But it was not missing from other textbooks! As usual, there was a lag in the publishing world, and throughout the 60's—to Flint's discomfort—other books continued to promote his now-discredited hypothesis.

By 1971, however, Flint could no longer ignore the Scablands and came around—in a grudging sort of way. In the third edition of his Pleistocene glaciation text, Flint yielded enough to devote one sentence to the Scablands floods: "Features, collectively known as channeled scabland, were widely created east of the Grand Coulee by overflow of an ice-margin lake upstream." This for what Baker rightly calls "the most spectacular event to occur during the Pleistocene on our planet"!

But nonetheless—shabby and belated as it was—Flint's concession marks a final victory. Ivy League authority had yielded;

J Harlen Bretz, about 94, seated in his favorite chair within reach of his favorite books at Boulderstrewn.

Bretz had won. Nearly 50 years had passed since Bretz first proposed the idea of catastrophic flooding, and now in 1971 (when Bretz was nearly 90 years old) his arguments had become a standard of geological thinking. It was a triumph of the highest degree, and it earned him at last, in 1979, the Penrose Medal of the Geological Society of America, the nation's highest geological award.

Bretz died on February 3 in 1981. He was 98 years old. Throughout his life he had been an active, independent thinker. He was guided by his own curiosity and sustained by tenacity and humor. Above all he was a teacher who left his mark not only on those who remember his classes with particular fondness but on the world as well. Bretz taught us all. He left us with a broader and a longer view, the same Olympian view that he himself had, so surprisingly, gained back when geology could walk but not yet fly, when it had boots but not yet wings. It was an extraordinary accomplishment, recognizing and promoting a geological story that no one wanted to hear. He was a man of both insight and courage, a modern-day Noah held by the vision of an unwelcome flood.

The story should perhaps end here, but true stories never do end neatly. They're too much a part of the whole. What Bretz began goes on, and even he is only a link between two ways of understanding. All things connect together. Early Catastrophism gave way to Uniformitarianism, which expanded our concept of time but stopped us from thinking in terms of sudden change. Then came Bretz's flood and with it still newer ways of perceiving, ways that include Uniformitarian ideas. You see, there's a great deal of sense in the idea of steady, imperceptible change. Most geology (from our limited, human viewpoint) works that way. And, the truth is, all those events we call "Catastrophic" are only the end result of slow processes building to rare but dramatic conclusions. Like the grand collapse of a carefully arranged row of dominoes, catastrophes are nothing more than finales.

In geology, believing is seeing. We're able to "see" catastrophes now. It was Bretz's Scabland that started the trend. Today there are a number of "catastrophic" hypotheses in the air. Alfred Wegener's long-maligned theory, leading to "plate tectonics," (with a story more intricate than Bretz's) has at last been taken to heart, and we can no longer consider the earth secure. The continents, we're told, are drifting, and the ocean basins are on the move. Africa and South America probably did once fit together, just as their jigsaw shapes suggest, and the West Coast of North America is apparently an alien piece of the Mid-Pacific that began to be shoved into the American continent back in the Mesozoic, 100 million years ago.

Nothing stays put. A mere 5 million years ago the dried-up basin of the Mediterranean Sea was cataclysmically filled when the Atlantic Ocean broke through a natural dam at Gibraltar. A tremendous volcanic eruption that occurred in 1470 B.C. on the island of Santorini may well have destroyed the legendary Atlantis. And now we hear that 50 million years ago star dust from a passing comet might have brought to a close the age of the dinosaurs. Where will it end?

Well, not in this world, at least. In the conclusion to his 1959 summary account of the Scablands, Bretz had written, "nowhere in the world is there known or suspected a story at all comparable to what we read from the scabland forms." And he was right, quite right; nowhere *in the world* is there a match for the colossal erosion and scarring of Eastern Washington. But there is *beyond* the world. In the 70's, when the Mariner 9 and Viking missions returned with their photographs of the Martian channels, geologists were confronted with scenes that resemble, more than anything, Bretz's Channeled Scablands.

It's all there, the anastomosing, the scouring, the bar formations. But why? And how? And from what? So the mystery goes on. We see bigger now, as Bretz taught us to, and think on grander scales. The Martian floodways are 60 miles across and have draining routes 600 miles long, and here we're dealing with a landscape 1 to 3 billion years old and 50 million miles from home.

It's always that way. Truth is never reached; it's only advanced. Other minds and other explorers take over for Bretz, nudging science along, bit by bit. Thesis, antithesis, synthesis, and modification of ideas. One thing leads to another, and the story never ends.

Part III: Setting the Stage

CHAPTER 12

GEOLOGY, LIFE AND THE FLOODS

Speak to the earth, and it shall teach thee.

JOB, 12:18

GEOLOGICAL FOUNDATIONS OF THE PACIFIC NORTHWEST

The Columbia River, seventh in length in this country and ninth in North America, drains more than a quarter million square miles of the Pacific Northwest. It drains over half of Oregon and Washington, nearly all of Idaho, a large area in northwestern Montana, and much of southern British Columbia. Its tributaries extend far into the rugged, glaciated Rocky Mountains of Canada. In their upper reaches, these rivers flow along down-faulted trenches within a rough terrain composed mostly of ancient rocks. Near the Canadian border, the Columbia is joined by the Clark Fork, an unusual river in that it flows westerly through Montana from the Continental Divide and drains the eastermost of these trenches.

The Columbia Plateau of Washington originated between 16 and 6 million years ago, when the greatest outpouring of lavas recorded in the history of North America spread out across 80 thousand square miles of western Idaho and eastern Washington and Oregon. The lava flows swept westward down a succession of broad valleys into the sea off northwestern Oregon, creating, from Astoria to Lincoln City, the rugged headlands which Oregonians call their own but which in fact emanated from fissures as far east as Idaho! During this time, the course of the ancestral Columbia River was repeatedly shouldered to the north and west by these enormous basalt floods, finally incising its present course near the northern edge of the flows. This up to two-mile-thick pile of black basalt, consisting of nearly 200 separate lava flows, is collectively called the *Columbia River Basalt Group*. But the flows with which we are concerned in this book are called the *Yakima Basalt*.

Weighted down by the over 90 thousand cubic miles of basalt, the earth's crust gradually sank for a period of 4 million years and, in doing so, produced the saucer-shaped Columbia Basin, which slopes inward from elevations of two to four thousand feet above sea level around the periphery to less than 500 feet at the lowest point in the Pasco Basin. About 12 million years ago, the pressures of this collapse began to warp the plateau into the numerous east-west sharp folds that characterize the western edge of the basin, just as the crust on a pudding wrinkles in cooling. Farther down the Columbia River in Oregon, the same processes of faulting and downwarping formed similar but smaller basins at the Dallas, Mosier, Hood River, and Portland.

Gravel, sand, and *silt,* carried by the Columbia River and its tributaries, began to fill these basins long before the outpourings of basalt ceased, so that we see successive lava flows separated by layers of basin sediment left behind by the river. Where these sedimentary layers occur between and above the basalt flows, in the Columbia Basin, they are called the *Ellensberg* and *Ringold Formations.* Those in The Dalles and Mosier Basins are referred to as *The Dalles Formation;* those in the Hood River and Portland Basins are known as the *Troutdale Formation.*

TIME AND ROCKS IN THE PACIFIC NORTHWEST

Age	Columbia River Plateau	Western Oregon
0–12,000 yrs.	Recent alluvium, talus, and landslide deposits	
12–15,000 yrs.	Touchet Beds and high gravel bars	Portland Gravels and Willamette Silts
15–30,000 yrs.	Older fluvio-glacial deposits	
30,000 yrs. to 1.8 million yrs.	Palouse Formation	Portland Hills Silt and terrace gravels
1.8 to 5 my.	Ringold Formation	Cascade, Boring and later lavas

Age	Columbia River Plateau	Western Oregon
4–13 my.	Ellensberg Formation and Yakima Basalt	The Dalles and Troutdale Formations
6–16 my.	Yakima Basalt of the Columbia River Basalt Group	

THE ICE AGE

During the long two million years of the last ice age, the great Continental Ice Sheet advanced from Canada into the northern United States and retreated again, four times. The periods of ice advance have been named the Nebraskan, Kansan, Illinoian, and Wisconsin glaciations, after good exposures of their deposits in those states. Each advance lasted from tens of thousands to a few hundred thousand years. During the melting and recession of the ice, long interglacial warm and dry periods intervened, some perhaps even warmer than at present. Records of the first three of these advances in the Northwest are scarce, but the record of the last glaciation, the Wisconsin, is abundant.

So much water was stored on land as ice during the glacial advances that the level of the sea was lowered by more than 300 feet, exposing wide coastal plains on the continental shelves of the world. The coastal plain off Oregon, for example, was at least 25 miles wide; off the Atlantic coast it was 150 miles wide. During the recurring warm interglacial periods, however, melting ice released its water and the sea repeatedly rose 150 feet or more above its present level. If this happened today all the major coastal cities of the world would be drowned! In the Pacific Northwest, as Thomas Condon suggested in 1871, high sea levels probably inundated the Willamette Valley with marine water during times of glacial melt, and the same would happen again if the ice sheets now covering Antartica and Greenland were to melt once more.

During the interglacial warm periods, the climate was so arid that the glacial outwash sediments in the valleys and basins along the course of the river (mostly rock ground up by the ice) were picked up and swirled about by violent dust storms all over the Northwest. In the Columbia Basin these deposits of wind-blown glacial dust and silt (loess) created the *Palouse Formation,* with deposits up to 150 feet in thickness. It forms the fertile farmland that makes the Columbia Plateau one of the great food-producing areas of the world, its deep, rich soil created by the

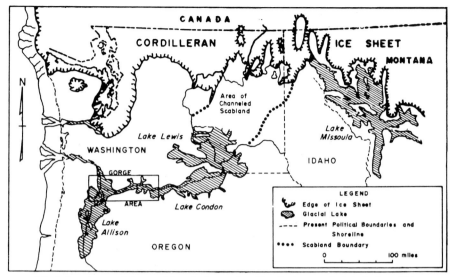

Glaciation, ice-age lakes and flooded areas (scabland when underlain by basalt) in the northwest. The maximum extent of the last ice sheet is shown by the hachured line. South of the ice, nearly 15,000 square miles was flooded or ponded in four temporary lakes: Missoula, Lewis, Condon and Allison. Note that the Pacific shoreline was many miles west of its present position, due to lowered sea level.

inhospitable agents of water, ice, and wind. Farther west, the uplands around Portland are mantled with similar loess, known as the *Portland Hills Silt*.

The last major advance of the Continental Ice Sheet began about 70,000 years ago in the Northwest, and it too consisted of several periods of glacial growth and retreat. Early progressions of ice probably resulted in large recurrent floods 60 and 50 thousand years ago. These were possibly even larger than the Bretz Floods which resulted from the very last advance of ice between 15 and 12.8 thousand years ago, which were of such a magnitude that most of the evidence left by the earlier floods has been destroyed.

THE PRE-FLOOD LANDSCAPE

What was the Pacific Northwest like 15,000 years ago, before the last floods? The hills and valleys were much as we see them today, except for the loess-covered Plateau and the V-shaped Columbia River Gorge. The peaks and uplands of the Cascades and the Northern Rockies were mantled with glaciers. Temperatures were 8° to 15° colder with 10 to 20 inches more rainfall than today, so that the overall climate resembled that of present-day Alaska. The rivers were larger; rainfall and snowfall were heavier. The Continental Ice Sheets covered much of Canada; south of the border, six lobes from the ice sheet extended for miles down the valleys of the Okanogon, San Poil, Columbia, Colville, Pend Oreille, and Priest Rivers in to Washington. Ice also capped much of the High Cascade Range and the Wallowa Mountains, and glaciers extended down the valleys, carving them into distinct U-shapes, like those now occupied by Lake Chelan in Washington and Wallowa Lake and the upper Sandy River Valley in Oregon. The Columbia Plateau was a great, undulating, gently-sloping, grassy, sparsely-wooded plain, bounded on the west by the Great Bend of the glacially-swollen Columbia River. Choked by glacial outwash, the river spread out in a multitude of channels across the Pasco Basin.

Westward, the Columbia River Valley through the Cascades and Coast Ranges was deeper than it is today. It occupied a narrow course in a steep-walled V-shaped valley, where cliffless walls sloped smoothly upwards to the adjacent summits.

In western Washington, the lobe of the Cordilleran Ice Sheet, which had been periodically scouring out Puget Sound for over a million years, had retreated to hover over Vancouver Island and the Straits of Jaun de Fuca, poised for its final southerly advance to Tacoma. The ice capping the Cascades and the majestic Olympics extended down the surrounding valleys and out into the Puget trough. Melt waters from these valley glaciers drained out to sea through the Chehalis Valley in a river much larger and deeper than the small, quiet stream we know today.

Today the coastal plains of Oregon and Washington are narrow strips of beach and terrace separated by lofty headlands; but in the period preceding the Bretz Floods, they formed one gentle westward slope for tens of miles to meet the lowered sea. This broad, fertile coastal plain, bounded by continual beaches and offshore bars and lagoons, would have furnished an easy travel route for any local or migrant tribes who lived off shellfish from the sea. Only the deep Columbia and Chehalis Rivers were there to prevent easy movement up and down the coast.

EARLY HUMANS IN NORTH AMERICA

But were there, in fact, people in the Northwest before the Bretz Floods? The question is still a controversial one. There is no undisputed evidence of the existence of humans in the Pacific Northwest before 12,000 years ago, near the end of the Bretz Floods.

When Europeans rediscovered the Americas in the decades following Columbus' first journey in 1492, they found two vast continents and associated islands, greater in area than Europe and Asia combined: lands extending from the arctic to the tropics and on southward to the fringes of the Antarctic; lush forests, jungles, grassy steppes and barren deserts; great rivers, high mountains and vast plains of the coast and the interior. Everywhere they found a balanced environment of land, sea and air and plants and animals.

Wherever the explorers went they found that other people had arrived earlier. They found nomads, fishers, hunters and gatherers. They found city dwellers and empires in Mexico and Peru. There were peoples of diverse cultures and tongues. They differed from group to group and were distinctly different from the Europeans. Some had special adaptations of the respiratory and circulatory systems which permitted them to live in the heights of the Andes or ill clad in the chilling winds of the Tierre del Fuego at the tip of South America. Europeans and American anthropologists later classified these many peoples into three large groups; the American Indians or Amerinds, the Eskimos, and the Aleuts. The Amerinds were by far the largest and most diverse group. They had settled in all parts of North and South America save for some of the arctic and subarctic areas occupied by small bands of Eskimos or Aleuts. The diversity of these peoples indicated that they must have been in the Americas a long time, indeed. But how long?

The early European settlers were certainly more concerned with their own survival than they were with the question of when the first people arrived in the New World. The settlers with a copy of the Bible may have responded by saying "4004 B.C. or later because, according to the Bible and Bishop Ussher, that's when everything began." Thus, for a couple of centuries it was generally accepted by laymen and scientists alike that human beings had been in the New World for less than 6000 years.

Then Charles Darwin published his *Origin of Species* (1859) and *The Descent of Man . . .* (1871). These works stimulated a world-wide interest in human origins. By 1892 Eugene Dubois presented the world with the skeletal remains of the Java Man and in 1927

Davidson Black discovered the Peking Man. These were primitive mid-Pleistocene people who had mastered fire for domestic purposes and who manufactured Old Stone Age tools more than 200,000 years ago. Did the descendants of these ancient people reach the New World? This is a good question. Let's explore it.

In the Old World there were human and human-like remains going back in older succession through the essentially human Cro-Magnon Man, the less human Neanderthal Man, the even less human *Homo erectus* (Peking Man) and finally to *Homo habilis*—the African handy man who used tools more than a million years ago. In parallel evolution there was a lineage of great apes or *Australopithecines* who roamed Africa and southern Asia. These appear to be near relatives of the early humans of the Old World. Are there similar parallels in the New World? No, there are none. Nowhere has any evidence of early humans other than *Homo sapiens* been found in the New World. *Homo sapiens* has no progenitors in the New World. It therefore follows that humans must have originated some other place and through wanderings came to and occupied the entire New World. How and when this happened is now becoming more a matter of scientific fact rather than conjecture.

It has long been recognized that Amerinds, Eskimos and Aleuts bear more resemblance to east Asians than to Europeans or Africans. Because of the differences between these racial groups, they have been named "Mongoloids" for the largest east Asian group; as "Caucasians" are named for the Europeans and west Asians; and "Negroid" for the large African group. The earlier assumption that all aboriginal Americans are Mongoloids who came from Asia has been recently confirmed by studies and research on comparative tooth configurations. It is now accepted by the scientific community that 1) early Americans came from Asia via Alaska, 2) the Amerinds came first and spread across most of the two great continents and 3) the Eskimos and the Aleuts came much later than the original Amerinds.

The remaining question is: When did the first successful Amerind settlers arrive and how long did it take them to spread acros North America, occupy the Carribean islands, cross the intercontinental bridge at the Isthmus of Panama and eventually people all of South America to the tip of Tierre del Fuego? This pioneering journey by a stone age people, a journey of 10,000 miles through forests, deserts, jungles, mountains and across daunting bodies of water: how long did it take? If the migratory pace were that of hunters, fishers, and gatherers, it may have been very slow, perhaps at the cautious rate of a mile a year—10,000 years. But if the migrants were hunting nomads in pursuit of big game (as some have proposed), the transit of the two continents may have taken only a few hundred years—along with the extinction of the elephants, mastodons, camels and most of the other large game animals of the New World. This riddle has not yet been unraveled but probably will be in the next few decades.

Like the geological controversy surrounding the Bretz Floods, the scientific community is split into two camps relative to the arrival of the Amerinds. There are those who believe in the Early Arrivers and those who believe in the Late Arrivers. The first school of thought holds that humans crossed the Bering Straits or the land bridge long before the last great advance of the continental ice sheet, 50,000 years ago. The latter group believe that the New World had no human inhabitants until about 12,000 years ago, when big game hunters, bringing with them the fluted Clovis spearpoint, proceeded in one great roundup to slaughter all the large Pleistocene mammals from Alaska to Patagonia.

The first migrations from Asia may have taken place across the Bering land bridge in early Wisconsin times, about 70,000 years ago; or they may have taken place during mid-Wisconsin times, between 20,000 and 60,000 years ago when the Bering Strait was open water but North America, except northern and eastern Canada, was free of ice. The Bering land bridge re-appeared during late Wisconsin, about 5,000 years ago, furnishing once again a dry crossing for humans as well as animals.

If our opinion is indeed confirmed, the early Americans who were unfortunate enough to witness or confront the Bretz Floods were descendants of migrants who arrived in North America more than 25,000 years ago. During late Wisconsin time, the period of maximum ice advance and the Bretz Floods, they were living south of the ice in what is now the United States and Mexico and north and west of the ice sheet in the Yukon. This means that some poor souls were living in the region devastated by the Bretz Floods.

A 1983 publication by Shutler and others describes numerous early American sites. Many of the sites described yielded evidence that the Amerinds occupied the North American continent prior to the Clovis point culture and the last advance of glacial ice over North America. Some of the dates are given below. The authors, compelled by this evidence, cast their vote with the school of Early Arrivers. The only direct evidence in favor of Early Arrivers in the region of the Bretz Floods, however, is a campsite with charred bones and stone artifacts buried under pre-flood deposits at the site of The Dalles Dam and a single stone artifact recovered from a Bretz Flood gravel bar at the mouth of the John Day River. This evidence was discovered by L. S. Cressman, an anthropologist now retired from the University of Oregon. It seems likely that additional and incontovertible proof of pre-flood human occupation of the Pacific Northwest will eventually be discovered.

Further evidence of human occupation in the Americas is also suggested by the following:

DATES AND SITES SUPPORTING THE OCCUPATION OF THE AMERICAS BY EARLY AMERICANS

Site	Location	Date B.P.*
Tepexpan	Mexico	27,000
Minnesota Lady	Minnesota	27,000
Midlands	Texas	27,000
Old Crow River	Yukon	27,000
Bluefish Cave II	Yukon	15,500
Bluefish Cave I	Yukon	12,900
Meadowcroft	Pennsylvania	19,000–12,600
Channel Islands	California	37,000–11,000
Manis	Washington	12,000
Laguna	California	17,150
Little Salt Springs	Florida	12,030
Fells Cave	Patagonia	11,000
Monte Verde	Chile	13,000
Los Toldos 3 Cave	Patagonia	12,500

1. **Actual remains:** At a site on Texas Street, San Diego, California, a human skull was found a number of years ago. It lay around in a museum for some time before it was dated (by means of a recently developed method called amino acid racemization) as being 48,000 years old. But the accuracy of the analysis, as in the case of other human fragments from at least 10 other sites in both Americas, has been questioned. In addition, carvings of extinct mammals on the thigh bone of an elephant were excavated near Puebla, Mexico and dated at 30,000 years ago.

2. **Authenticated artifacts of bone and stone:** A few more sites have been found that contain artifacts, but the dating of these sites have also been under criticism.

3. **Possible human sites:** Dozens of sites older than 10,000 years contain charcoal, red clay hearths, and pieces of bone and rocks that might have had human use. The dates of 30,000, 40,000

*Before Present

and as much as 65,000 years from these sites are valid, but the actual presence of humans has not been authenticated. The famous Santa Rosa Island site, where bones of an extinct pygmy elephant were found with crude artifacts, has been dated at 30,000 years ago. Yet this has also been doubted.

4. **Population growth and travel time:** Estimates of the population of the two Americas in 1492 range from 8 to 75 million (Kroeber, 1934). Population pressure is one of the factors that determines rates of migration, the other being availability of game animals for a hunting culture. "It took much time ... many millennia—to breed so many men and to cover so wide a space" (MacGowan and Hester, 1962). The distance from the Bering Straits, where humans presumably first migrated to the Americas to Cape Horn at the southern tip of South America is 10,000 miles.

5. **Diversity of tongues:** There are at least 160 different linguistic stocks or language families in the two Americas, more than in all the rest of the world (Kroeber, 1934). "It would have taken at least 20,000 years, perhaps three times that, for such diversity to have developed" (Harrington, 1940). "There is no way that this diversity could have developed in less than 60–80,000 years" (Pierce, 1982).

The accumulation of so many suggestive data from a number of completely different lines of evidence is enough to convince the writers that people came across the Bering Strait before the last advance of ice during a time when the oceans were at their lowest stage and there was a bridge 1000 miles wide, now occupied by the Bering Sea.

LIFE BEFORE THE FLOODS

What would the world have been like for these people? West of the Cascades, on the wide plains of what is today the Willamette Valley, enormous herds of great mammals ranged. Mammoth and horse (both doomed to extinction—the mammoth a little more than 10,000 years ago, the horse a little more than 6,000) were especially abundant along the meadering river bottoms and across the prairies. But roaming the wide grasslands of eastern Washington and Oregon, the herd animals of the late Ice Age were more varied and plentiful. Wooly mammoth and mastodon, longhorn bison, camel, caribou, and musk ox roamed the plains. Close behind, and preying on these herbivores, came the hunters, wolves in packs or lone saber-toothed tigers hunting by stealth. Giant condors soared across the skies, in search of carrion left by the hunters; and along tributaries to the main rivers, giant beavers built dams, while shaggy, short-faced bears vied with early humans for the abundant salmon that swarmed up the Columbia and its side streams to spawn in the shallows.

Bighorn sheep roamed next to the glaciers on the slopes of the Rocky, Blue, and Wallowa Mountains. Lower on the pine-covered slopes, deer and camel browsed on shrubbery and low-hanging branches of aspen and oak, maple and alder.

South of the Blue Mountains in eastern Oregon, numerous broad shallow lakes filled the basins; pelicans dived in the fish-filled waters; and the skies, were darkened by the seasonal migrations of countless ducks and geese. The abundant fish, birds, wildlife, camas root and wapato (an edible potato-like root), made life relatively comfortable for early humans living in the area. There were also lake reeds to be woven into sandals, mats, and baskets. It was a rich world for the small wandering bands (made up of perhaps 10–12 individuals) living by hunting and gathering. It was, as well, a world with considerable solitude and space.

Today, even in the heart of what we think of as wilderness, civilization intrudes. The sky is invaded by aircraft, marked by trails of passing jets. Motor noises, human voices, and scraps of discarded human belongings crop up in the most remote localities. But before 15,000 years ago, in the late Pleistocene, isolation was the rule. The sounds of one's own band, with its chatter and activity, was the only human sound to greet the hunter or gatherer returning to the encampment. It was a world far larger and far more powerful than the people who lived within it, a world both bountiful and foreboding.

THE FLOODS

"Imagine the kind of velocity the water had when the flood came through, the tremendous air blast caused by the wave front. Its roar would have built for a half-hour at least, arriving as a series of pulses.

Following on the great wind came the crashing turmoil of water, piling onto itself in the breakthroughs. Had the observers been at high water level, they might have been able to run a few hundred feet for their lives."

Leonard Palmer, *Northwest Magazine*, June 26, 1983

This is how we must conceive of a Bretz Flood if we wish to see it through the eyes of those people who lived in the Northwest 15,000 years ago. With some considerable stretching of the imagination, we can conjure up a vision of how these floods must have appeared, both to the few who survived it and to those others who knew it only in a brief and terrifying final consciousness.

The roar, faint at first but quickly increasing in intensity, would have been audible for at least the half-hour before the flood struck. Who, among those who heard the noise, could have imagined what it meant or conceived a plan of action?

At the crescendo came destruction. The mass of compressed air—impelled by the towering head of onrushing water—hit first. We know about winds building over time into hurricane strength, but how can we imagine a torrent of air exploding into existence, driven by a wall of water hundreds of feet high and moving at 50 miles per hour? Think of shock waves from a volcanic eruption or an atomic explosion but think of these shock waves as emanating from the face of one plunging, thundering, rolling water-wall, moving at speeds only slightly less than that of a car on a freeway, a swollen, surging mass composed of ice, rock, mud, and water, and standing over 500 feet tall for the first 100 miles of its discharge, and again this tall where the waters were compressed and funneled by constricting features such as the Wallula Gap and the Columbia River Gorge. Think too that, for some, this wall may well have come at night.

How can we conceive of the force contained in this water mass? Imagine water blasted from a fire hose, but think of this hose has having a nozzle as wide and tall as the valleys or channels through which the flood burst. The torrent, rushing from Lake Missoula down through the Wallula gap, ravaged and swallowed everything that lay in its path—a cornucopia gone mad, pouring forth destruction.

It is no surprise that signs of human occupation in the river basin older than 12,000 years are missing (with the exception of a few possible artifacts found within the gravels). The flooding would be most concentrated and most destructive along the valley bottoms, in the very places people were particularly likely to congregate. We can only imagine what sort of legends and taboos were created among peoples who depended upon the river for their existence and yet were periodically destroyed by catastrophic flooding!

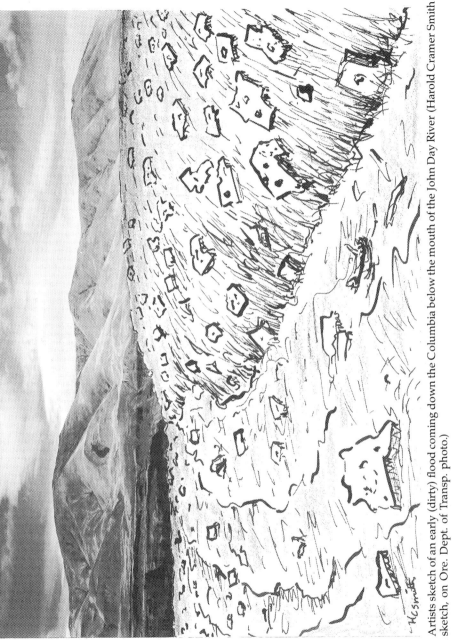

Artists sketch of an early (dirty) flood coming down the Columbia below the mouth of the John Day River (Harold Cramer Smith sketch, on Ore. Dept. of Transp. photo.)

CHAPTER 13

INTRODUCTION TO CATACLYSMS

Yon foaming flood seems motionless as ice;
Its dizzy turbulence eludes the eye,
Frozen by distance.
WORDSWORTH *(Address to Kilchurn Castle)*

After the floods, 16,000 miles of grassland, plains, and river valleys were drastically changed. We have attempted to imagine the floods themselves. The next effort of the imagination is to envision the landscape in the days and years that followed the devastation. What records or indications do we have? Those few individuals who survived the deluges have themselves been dead now for 12,000 years. Having no written language, they left no written account; and though there are various flood legends among the stories of the Native American people, none suggest the swift destruction and the awesome rushing magnitude of the Bretz Floods.

We are working with an event so long in the past and so foreign to our experience that description of any sort is inevitably less than adequate. Probably there is no way we can create a real picture of the floods and what the landscape looked like after the waters receded. It's like reading about the effects of nuclear war— inconceivable and therefore too easily believed impossible. Perhaps the best we can do is to look to the closest and largest geological events of our own times and draw comparisons.

For those of us in the Northwest, the most immediate and memorable geological catastrophe is the 1980 explosion of Mount St. Helens. Here, in a sense, was a disastrous natural phenomenon we all experienced. Television, more than anything, brought it close; and photographs, publications, and personal accounts have told us again and again the story of the mountain's eruption. But we learn that the eruption, meant different things to different individuals. Those upwind from the ash who witnessed it from a safe distance were, for the most part, excited and inspired by the eruption. For others, however, the effects were more immediate: troubling, terrifying, or disastrous and, in some cases, tragic.

Even today it is impossible to see the area of devastation that spreads out to the north and west of the mountain without being deeply affected. President Carter called it "the worst thing" he had ever seen. "Absolute and total devastation of an area that encompasses about one hundred and fifty miles . . . much worse than the description which had been impressed upon me . . . much worse than anything I've seen in pictures of the moon's surface."

Still, even from the air (it was from this vantage point that President Carter saw the devastation), comprehension remains difficult. The landscape cheats the eye; its scope evades us. Only when another airplane moves across our vision five miles or so away and shows itself mosquito-sized against the mountain's broken flanks or within the crater rim does the scene suddenly shift and swell and for a moment reveal its true and gigantic proportions.

But the Mount St. Helens eruption was not a Bretz Flood. It is merely something we can refer to as a way of approaching the earlier event. The fact is, Mount St. Helens produced a paltry little disaster compared to the devastation of the Bretz Floods, but it can serve as a model to help us gain an insight into the larger, more disastrous Pleistocene event, just as Bretz's familiarity with Puget Sound glacial flooding prepared him to envision flooding on an even larger scale. Mount St. Helens was an explosion. The Bretz Floods were massive deluges. Mount St. Helens' ash lies spread mantle-like over hills, ridges, and slopes, varying from a thin trace to layers several feet thick. It was blasted to where it lies or carried by winds until it descended to the earth from above. But the Bretz Floods were earth bound and rushed over and through the land. The immediate effects were scouring and depositing of debris at a level no higher than the waters themselves. Hill tops above the flood remained untouched.

In the early days after the Mount St. Helens eruption, boulders and ice blocks lay scattered over—and embedded within—the new valley floor to the north and west of the mountain. Some of these ice chunks, as large as small houses, moved down the valley, caught in the mass of descending debris. Others, as large as small cars, had been blasted through the air as far as 8 miles from the mountain's crater. In time, the ice melted; for some of the larger blocks, the melting took months. And where such ice-boulders have vanished, hollows are left behind in the already lumpy and pock-marked terrain. Some of the meltings formed pools; others drained more quickly or evaporated.

There were other effects as well. Where slurries of mud, ice, and debris rolled down the river valleys, new drainage routes in characteristic braided patterns were cut through the valley floor (now up to 300 feet above the buried pre-eruption channel). And in the early months, buried, heated rocks vented steam through the thick coating of mud and ash, creating a particularly eerie and hellish effect.

Even today there's a dump-ground look to the place, a raw and wounded look. Though nature has started to heal the earth, though some hints of green have begun to relieve the putty-gray on the north side of the mountain, recovery is slow and the scene— from the air or ground—will remain broken and depressing for years to come.

With the Bretz Floods it was all a bashing and grinding of water and ice and stone. It was flooding on a more than grand scale (much greater than the river-valley flooding that Mount St. Helens produced on the side). The effects of the massive Bretz floods were both extensive and enduring. Certainly the water crest must have subsided within a day or two, but the remaining discharge may have taken weeks to drain away. And since some of the water was inevitably trapped, a period of lakes, ponds, and puddles followed the floods themselves. This occurred for two reasons. For one, the impact and scouring of the rushing, swirling water-mass gouged out depressions; and here, within these newly-created hollows, water was trapped and remained so until—with time—drainage, evaporation, or sediment fill-in changed the land.

But the floods also had another means of creating lakes and ponds. Consider what must have happened when vast amounts of debris swept down the floodway, through the Scablands, down the Columbia Gorge and into the valleys below. The strong current in the middle would flush the debris on through. On the sides, however, on the edges and in the backwaters, were slower currents; and here the debris dropped out and settled in the piles Bretz began to recognize as gravel bars. As these bars grew higher with the passing of the flood, they formed hill-sized mounds that created natural levees. These filled up the mouths of side valleys and barricaded drainage routes. Then, once the main floodwaters receded, a series of lakes came into existence within these now-barricaded side valleys. When a lake level topped its debris dam, it rapidly cut through the material, breaching the barricade and quickly draining the lake. In some cases, however, the blockage of the lakes or ponds might be maintained for months or years.

When we try to imagine what these lakes must have looked like—both those lakes formed from blockage and those lying within newly-created hollows—, we can turn again to the river channels of Mount St. Helens. Along the drainage routes running off the mountain, debris piles and blockages similar to (though much smaller than!) those caused by the Bretz Floods developed. The dump-ground image is valid for both catastrophes. Both formed a series of various-sized ponds and lakes lying at various levels like flat, random steps set within a rubble field. On Mount St. Helens' recreated Spirit Lake, float large numbers of blasted trees, moving from one end of the lake to the other as the wind shifts. So it was in the lakes formed by the Bretz Floods, save that the floating objects were mostly icebergs drifting from one end of a lake to the other.

But then, of course, these icebergs would melt. Their cargo of rocks and solid debris would drop out, float off or sink to the bottom of the lake. In this way the Floods scattered their erratic boulders, and in doing so they created an excellent gauge for recording lake levels. In the Willamette Valley, for example, the greatest number of erratics are found below the 400 foot level, indicating that the waters went no higher and that a large number of the drifting bergs were grounded in the shallows at the lake edges. Remember now that lakes, like waterfalls, are mere geological accidents and are temporary features. Most of these flood-produced lakes drained or evaporated away within months, years, or decades; but some lasted longer. The gravel bar formed behind Rocky Butte in Portland, for example, appears to have created a lake extending 70 miles, as far south as Albany. Here was a lake larger than the San Francisco Bay, and it remained in existence long enough to form river deltas and beach notches, where the waves cut into side hills at the 200 foot level. Today, of course, this lake no longer exists, but it has left behind shoreline marks on the hills—traces recognizable to the geologically initiated.

At first, then, there was a period of slowly disappearing lakes and ponds, with the receding waters leaving their scattering of erratics here and there on the lake floors, as well as in a bathtub ring at the high water levels. But what happens once the water is gone?

Remember that the Bretz Floods were turbulent floods, full of sediment and debris moving in a vast, thick, muddy slurry. As the waters slowed, this debris settled out, leaving the cleaner, upper waters to drain off. What is left is mud, and mud dries into dust, and dust is unstable. When winds whip across such stretches of dried, dusty soil, clouds of the material are taken up, carried into the atmosphere, and deposited over the countryside. The greatest amounts, of course, settled nearby. The West Hills of Portland were in part covered this way, when the Bretz Flood muds dried out and blew up from the Willamette Valley to collect on hillsides and hilltops. Here, above the crest of the Floods, the wind-borne soil largely remained where it landed, anchored by plants.

But some of the dust went farther. Like the Dust Bowl storms of the 20's, some of the Bretz Flood dust must have also moved great distances through the air. Certainly some must have fallen on the East Coast or even further away. We know, for example, that people in Massachusetts were able to collect Mount St. Helens ash by spreading aluminum foil to gather the fine grains falling from the air.

Rains would, of course, settle such dust storms, regenerating mud. Yet, after a time, plants would return, sink their roots in, spread their leaves, and hold the dust in place. We can see this happening on Mount St. Helens. In a similar way it must have happened along the dusty route of the Bretz Floods. With time, the ravages have been softened and covered. Today, only the raw scars of the Channeled Scablands clearly show the devastation that passed over our land centuries ago.

These, then, are the stages in the recovery following the Bretz Floods as we can best conceive of them. But it is important to realize that these "stages" are only stopped frames in a continually changing process. We should attempt to see the Floods and their after-effects as a flow of events: glaciation, floods, lakes, dust, and the re-establishment of plant and animal life. And all these interconnecting stages have left their marks (spoors and tracks; gravel bars and bathtub rings), the very clues that geologists look to in their discovery of the earth's history.

CHAPTER 14

AFTER THE BRETZ FLOODS

. . . and the waters were abated from off the earth . . .
GENESIS, 8:11

THE LANDSCAPE

Just what was the Pacific Northwest like 12 thousand years ago, *after* the last flood? As the melting ice fronts in the intermontane valleys of the Rocky Mountains retreated, lakes formed behind the terminal moraines, those piles of earth and rock debris that are deposited in front of an advancing glacier. The largest of these lakes is now known as Flathead Lake in Montana.

Dozens of other lakes in the trenches became filled with glacial outwash, became swamps, and were eventually cut into by streams to form terraces along the valleys and new plains. In the Purcell Trench, lakes Couer d'Alene, Pend Oreille and still farther north Kootenai Lake were larger than now; they since have been reduced in size by deltas where entering streams deposited their load.

The once smooth rolling plains of loess-covered basalt on the Columbia Plateau were now deeply scarred into hundreds of miles of broad and narrow dry channels and deeper coulees. Their scabland bottoms were dotted with kolk lakes whose basins had been plucked out in the barren basalt, particularly in the Cheney-Palouse area in the east, and the Grand Coulee to the west.

East of the Cascade Range, the great lakes that had covered so much of the southeast part of Oregon slowly wasted away in the newly semi-arid climate. Lower and lower wave-cut terraces still mark their ice age shorelines. By 8000 years ago, the continued warming and drying of the oncoming Altithermal (warmer climatic) period had begun to produce a desert much like present-day Nevada; and even lakes such as Summer, Abert, Klamath, Hart and Goose were smaller than now or had disappeared entirely.

The Columbia River tooks its present course, much reduced in volume but still powerful enough to clear accumulated glacial outwash from the upper reaches of its channel. The Yakima and Walla Walla Rivers and their tributaries began to cut rapidly down through the soft layered flood gravels and silts that had filled their valleys, leaving the flat-topped benches and terraces that now lie well above the valley floors.

In the Umatilla Basin, the north-flowing tributaries of Fourmile, Sixmile, Eightmile, Pine, Willow and Coyote Creeks began again to cut their straight shallow canyons through the wide expanse of flood gravels. Below The Dalles, the flood-deepened channel of the Columbia coursed below steep and barren cliffs, some 1000 foot high and nearly vertical on the south side. The flood-widened valley, with its new spectacular tributary waterfalls, now began to fill with gravels from upstream as sea level rose.

In the Portland area and in the Willamette Valley to the south, where rainfall continued to be greater than east of the Cascades, the numerous small lakes produced by scouring began to fill slowly. Former Lake Cipole (now Onion Flat) west of Tualatin, and Lake Labish north of Salem are now peat-filled flats. Others, such as Lake Oswego, Lakamas Lake north of Camas, Washington, and lakes in the valley of Salmon Creek north of Vancouver, still remain.

Between Portland and the sea, the rising sea level inundated the over 300-foot-deep Ice Age canyon and the coastal plains which had extended for 25 miles west of Astoria. During the Altithermal period, 8000 to 4000 years ago, sea level stood a few tens of feet higher than today, and waves washed against sea-cliffs east of present Highway 101 between Warrenton and Seaside, and between Tillamook and Nehalem Bays.

North of the mouth of the Columbia River, the sea covered nearly all of Long Beach peninsula west of Willipa Bay. Farther north, the great ice age Chehalis River, which drained the Puget Sound for the tens of thousands of years that ice covered the Straits of Juan de Fuca, became the small meandering stream of today.

THE LIFE

Within a few hundred years, the warming climate encouraged pine and juniper to recover on the drowned walls of the Clark Fork valley within the area covered by Lake Missoula.

Lush grasses again covered the silt-filled basins and upland islands of loessal silt which remained on the Columbia Plateau in eastern Oregon and Washington, but never again covered the stripped scabland areas of bare basaltic rock. Along the Columbia River Gorge the Douglas fir, hemlock, cottonwood, big leaf and vine maple began again to take hold, after the first growth of fireweed, alder and other underbrush prepared the way.

The time of the "great extinction" of most of the huge mammals which were dominant during the Ice Age began soon after the floods ceased and continued for about 7000 years. The wooly mammoth and the long-haired bison went first, 11 thousand years ago, followed in a few thousand years by the musk ox, the horse and the mastodon. The camel lasted until only 3000 years ago.

While not yet certain, most scientists believe that Amerind hunters were the chief agent of these extinctions. The abrupt change in climate which dried up the lakes and changed the vegetation surely contributed to their decline, and the rise of sea level which flooded the formerly broad coastal plains certainly reduced their habitat by many thousands of square miles.

The Indians returned to their old fishing grounds, and found that the new channels and rapids formed at Celilo by the floods made ideal places for catching salmon. It was for this reason that Celilo became an annual meeting place for tribes from many miles around.

CRITERIA FOR THE RECOGNITION
OF PAST GREAT FLOODS

If the velocity of a stream is doubled, its ability to move material along the bottom increases up to 64 times, and its ability to carry finer material in suspension also greatly increases. Great floods thus move fantastic amounts of earth and rocks, which are then deposited in or along the channel farther down the valley, when the current slackens.

It is estimated that one annual flood transports at least 90 percent of the earth and rocks moved along the bed of a normal stream during an entire year. A river flowing with a velocity of two feet per second (1.3 miles per hour) may speed up to 5 feet per second (3.4 miles per hour) during an annual flood. This increases the amount of material moved by a factor of nearly 40 times. Exceptionally large floods may move up to 10 feet per second (nearly 7 miles per hour).

The Bretz floods are estimated to have had velocities in

some places of as much as 75 feet per second (50 miles per hour), which would increase their capacity for moving material by many thousands of times.

Both erosion and deposition of earth materials by abnormally great floods leave clear and long-lasting evidence of their passage. Twelve unique landforms and aspects of sediments left by the Bretz floods may be observed throughout over 16,000 square miles in the Northwest.

EROSIONAL FEATURES

1. Relative scarcity of soil or soft sediment along floodways below the level of the highest flood crest. Soils may take more than 10,000 years to reform!

2. Prominent high-water marks and shoreline features, cut into soil or rocks or deposited by waves and currents at heights hundreds of feet above present river levels.

3. "Scabland" topography (mesas and intervening dry channels) resulting from the scouring and irregular plucking out of pieces of rock (fractured joint blocks) from the underlying Yakima basalts.

4. A braided or "anastamosed" pattern of the numerous dry channels, as well as divide-crossings where the floodwaters top their channels and spill over onto other channels or valleys producing kolks, or "closed depressions" (hollows with no drainage outlets) in the channel floors.

5. Widening and deepening of main valleys, forming steepened walls and cliffs (escarpments), faceted spurs (where a ridge, sloping down to the valley has had its lower end cut off) and hanging valleys (where the lower part has been cut away to form a waterfall) at the mouths of tributary streams.

6. Landslides or potential landslide areas, resulting from undercutting of valley walls or saturation of underlying soft sediments by water.

DEPOSITIONAL FEATURES

7. Presence of ice- (and tree-) rafted erratic boulders and smaller fragments of granite, schist, quartzite and other rocks which are not part of the natural rock formations where they are found.

8. Presence of originally round pebbles that have been broken into angular shapes, cracked by impact in high-energy torrents.

9. Longitudinal or eddy bars, perched high on the walls of the flood-scoured valleys. These are found below (downstream from) promontories or in reentrants (recesses or side valleys). Pendant bars are also common; they extend downstream behind the protection of a rock island or other such obstruction in the path of the floods. Expansion bars form where the water spreads out after passing through narrows and deposits some of its load.

10. Giant ripple ridges up to 50 feet high and 500 feet apart on the surface of both expansion and longitudinal bars below narrows. These "giant ripple marks" require currents of up to 50 miles per hour and depths of up to 500 feet for their formation!

11. Foreset bedding in gravel bars. Gravel bars build and move downstream during floods by the top material rolling downstream in the current and spilling over the steep lower end of the bar. Eventually, each bar is thus composed of beds that slope steeply in a downstream direction. These foreset beds are found along the sides of canyons tributary to the main river, and they dip in an opposite direction from what is normally the tributary stream's main current; in other words, they dip upstream, a clear indication that flood waters rushed up these tributaries and dumped part of their load.

12. Rhythmite sequences in tributary streams, composed of many layers, each one consisting of gravel or sand at the base, followed by silt and then clay at the top. Each rhythmite represents a separate flooding of the tributary stream. Within some of the rhythmites are sequences of annual layers called *varves*. These are much finer-grained, consisting of fine silt and clay deposited in a temporary lake in a side valley. By counting the number of varves between the larger and coarser flood deposits, one can estimate the number of years between successive floods.

Part IV:
Following the Floods
From Source to Sea

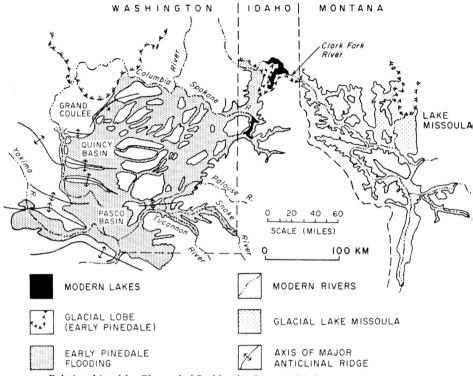

Relationship of the Channeled Scablands of eastern Washington to Lake Missoula (after Baker, 1973).

CHAPTER 15

MISSOULA, MONTANA, TO SPOKANE, WASHINGTON

The rock scourings are the trails left by the invader. Their character should reveal the nature of the icy visitant as tracks reveal the track maker.

T. C. CHAMBERLAIN, *1888*

INTRODUCTION

Most geologists now believe that for slightly more than 2000 years (a period roughly the same as from the glorious period of Athens to the present), at least 40 tremendous cataclysms of almost inconceivable force and dimensions swept across large parts of the Columbia River drainage basin. These occurred between 15,000 and 12,800 years ago, at the end of the Ice Age.

More than 16,000 thousand square miles were repeatedly inundated to depths of hundreds of feet, and the landscape was transformed by the greatest scientifically documented floods known to have occurred in North America. These floods on the Columbia contained up to ten times the flow of all the rivers in the world, 60 times the flow of the Amazon River.

The first floods were undoubtedly the largest and "dirtiest" of all, carrying innumerable rock-laden icebergs from the ice dam and accumulating vast amounts of soil and rock torn from its bottom. The frontal margin of such a flood must have resembled a slurry or mudflow, rather than a watery deluge. Not only did the floodwaters spread out to form multitudes of channels by stripping off all vegetation and up to 150 feet thicknesses of cover from the loess-mantled plateau, but they also picked up stream-bottom sediments in the old stream valleys and bit deeply into the underlying lava flows to produce the hundreds of so-called kolk lakes which now dot the dry courses of the channels. Rock Lake, south of Spokane, one of the largest of these, is over 7 miles long and approaches a mile in width.

As they swept across the Columbia Basin, nearly 50 cubic miles of soft silt, sediment and hard lava were carved out of the earth's skin to form the network of scabland channels, whose eroded basalt surfaces and dry falls now typify large parts of the Plateau. Downstream in Oregon, the valley and Gorge of the Columbia was scoured out all the way to the sea. The Willamette Valley was flooded all the way to Eugene.

Thanks to J Harlen Bretz, the cause of the deluges is now understood and amply recorded in the technical literature, though relatively little has been published on the results of the flooding in Oregon.

It happened this way. The lobe of the Cordilleran ice sheet that occupied the Purcell Trench in British Columbia advanced southward down the trench to and beyond Pend Oreille Lake. Each time it advanced up the Clark Fork several miles it formed an ice dam as much as 2500 feet high across the valley, impounding the waters behind the dam to form a great lake up to 2000 feet deep, covering 3000 square miles, and extending for 200 miles to the east in the intermontane valleys within the Rocky Mountains.

Each of this series of pre-historic lakes, now known collectively as *Lake Missoula,* contained over 500 cubic miles of water, one-fifth the volume of Lake Michigan. When the rising waters became deep enough to float the ice, they lifted up the dam and the ice was swept away. Within a few hours or days, up to 380 cubic miles of ice-choked water surged out at an estimated rate of 9.5 cubic miles an hour and swept southwesterly at speeds of from 30 to 50 miles an hour across the Columbia Plateau.

Each time Lake Missoula emptied, the ice lobe, continuing its southerly progression, would build a new dam and form a new lake, resulting in a new flood. This happened on an average of every 55 years or so for 2000 years!

LAKE MISSOULA

(Montana Highway 200, U.S. 90)

Lake Missoula flooded the Clark Fork Valley along U.S. 90 all the way east to beyond Drummond, within a few miles of Deer Lodge, and along the Thompson, Little Bitterroot and Flathead (Mission) Valleys to the north beyond Kalispell (U.S. 93), the Blackfoot and Nevada River Valleys to the north and the broad Bitteroot Valley along U.S. 93 to many miles south of Hamilton. The site of the city of Missoula lay under 950 feet of water, as can be seen by the shoreline markings on the hill east of town. The water reached

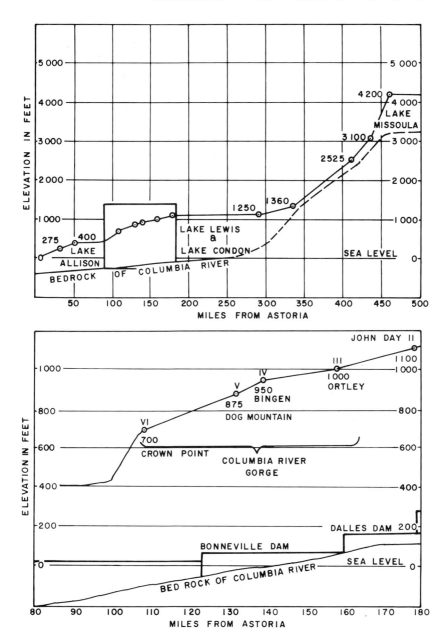

(Upper) Longitudinal profile showing the maximum elevation of the Bretz floods from Lake Missoula to the sea, and bedrock elevations along the course of the floods. Square shows location of the figure below. Miles are from Astoria. (In part after Richmond, et al., 1965)

(Lower) Longitudinal profile in the Columbia River Gorge, showing the maximum elevation of the flood crests at measured cross-sections (II to VI), and elevations of modern dam pools and bedrock. Miles are from Astoria.

depths of 1100 to 1400 feet in several of the other basins to the west and northwest.

In the northern part of Camas Prairie Basin, a few miles north of Perma (Highway 200), an area of over six square miles is covered by hundreds of parallel gravel ridges and bars lying upon bedrock that Pardee called "giant ripple marks." They are so large (20 to 30 feet high and 200 to 500 feet apart) that, as he said, "the term ripple mark seems inappropriate." They can be seen in and below lower points in the Markel Pass divide (Highways 28 and 382) south of Camas Hot Springs. Some are nearly two miles long. They could only have been formed by floods at least 800 feet deep plunging across the pass at velocities up to 55 miles per hour!

From 35 to 40 rhythmites in Lake Missoula sediments have been described by Waitt, 1977, 1978, Alt and Chambers, 1970, Chambers, 1971 and Curry, 1977. They interpret these as representing the same number of successive fillings and drainings of the lake. Waitt (1980) proposed that each lake rose for about 30 to 60 years before breaking through and defeating the ice, by floating away the dam with all its contained debris and sending it seaward in a catastrophic surge. Then, as the ice lobe continued its advance down the trench, another ice dam was formed. Dam after dam formed and lake after lake was filled by the heavy rainfall, snowfall, and melting glacier fronts in the trenches far to the east. Each dam was successively swept away by the impounded waters.

The actual number of floodings remains a controversial issue. Geologists have proposed as few as 2 and as many as 70 floods. The numerous rhythmites, believed to indicate many individual floodings, are interpreted by others as representing varying lake levels during the periods between floodings, or successive surges during individual floods. Geologists, however, are used to working with a wide range of possibilities. The large number of independently variable factors in the discipline requires geologists to use what is called the "method of multiple working hypotheses" (Chamberlain, 1897). This consists of devising as many explanations as possible for the relationships between the groups of data and then using field and laboratory work and knowledge of geological and climatic conditions to test them. A geologist will selectively abandon the least possible and retain the simplest and most plausible.

Here, for example, is one way of testing the possiblity of multiple flooding. First consider how long it would take the 500 cubic miles of water impounded in Lake Missoula to be replenished in 30–60 years (the amount of time Waitt suggested for each filling). The present annual flow of the Clark Fork River

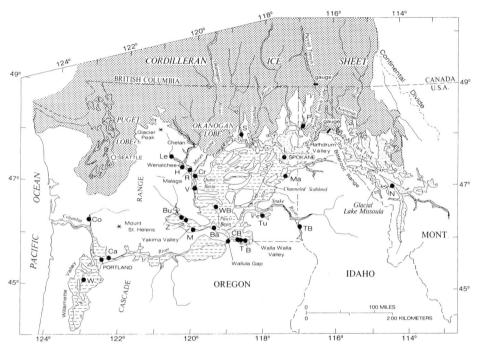

Map of Columbia River valley and tributaries. Small dots show maximum area of glacial Lake Missoula east of Purcell Trench ice lobe and maximum extent of glacial Lake Columbia east of Okanogan lobe. Lined pattern shows area that, besides these lakes, was swept by the Missoula floods. Closed circles indicate sites of bedded flood sediment discussed in text: Bu = Buena; Ca = Camas; Co = Cowlitz valley; Cr = Crescent Bar; CB = Cumings Bridge; CR = Castle Rock; H = Horse Lake Canyon; L = Latah Creek; Le = Leavenworth; M = Mabton; Ma = Malden; N = Ninemile Creek; P = Priest valley; R = Rock Island bar; V = Vantage; W = Willamette valley section; WB = White Bluffs; Z = Zillah.

would be sufficient to fill the lake in about 135 years. But an ice age lake could have filled in less than half that time because the climate at that time is believed to have been colder and wetter than it is today. Rainfall was 10 to 20 inches greater, perhaps 36 inches in northern Montana. Evapo-transpiration for that area today is about 5 inches; it must have been less during the Ice Age. So, a source of water to fill the lake in 60 or fewer years can be identified.

Rhythmites on U.S. 90, 7 miles east of Alberton, Montana. Each layer may represent a separate Bretz flood (Carson photo).

PEND OREILLE TO SPOKANE

(U.S. 2, 90, 95 and Idaho 41, 53)

When the waters in Lake Missoula broke through the 2500-foot high ice dam just southeast of Lake Pend Oreille (U.S. 2), they poured south for 20 miles, and then southwest (U.S. 95) along the Purcell Trench across Rathdrum Prairie into the narrows west of Post Falls and down the Spokane River Valley (U.S. 90-10 and Idaho 53) for 40 miles to Spokane. The maximum flow was more than 9.5 cubic miles of water per hour, which could have drained the lake in two days. More probably it slowed and lasted for a week or more. Along this 70-mile stretch, the crest of the flood dropped from 4200 to 2650 feet at Rathdrum Prairie, with a gradient steeper than anywhere except at cataracts, such as Grand Coulee. Velocities may well have reached as much as 50 or 60 miles per hour east of Spokane. The water in Lake Couer D'Alene barely spilled over the divide between Lake and Rock Creeks (Washington 58).

Giant ripple ridges were formed on Rathdrum Prairie east of Spirit Lake. Gravel fill from the floods may have helped plug the mouths of tributary streams from the west to form Spirit and Twin Lakes (Idaho 41), and Hauser and Newman Lakes (Idaho 53). To the east and south, flood sediments may have contributed to the damming of Hayden, Couer D'Alene and Liberty Lakes (U.S. 90-10).

The floods were as much as 500 feet deep above the present site of Spokane. The waters breached the wide low divide into the upper reaches of the Cheney-Palouse scablands to the south, and also poured northwest down the Spokane River into the Columbia towards Grand Coulee.

Just south of Spokane on the east bank of Hangman (Latah) Creek, 3 miles south of its junction with the Spokane River, Rigby (1982) recently described a sequence, lying below 1800 feet in elevation, of up to 20 rhythmites that were deposited in the ponded water of that side valley. The rhythmites are from 3 to 17 feet thick. Each of them is composed of a lower gravelly portion and an upper portion of fine silt and clay (laminated into 10 to 60 fine layers with some of the upper rhythmites). The thin laminated layers are interpreted by Rigby as being annual varves. If he is correct, at least 20 separate floods have left their record at Hangmans Lake. Each of these floods was separated by 10 to 60 years.

Closer view of Latah Creek exposure, showing the character of the rhythmites. Note large number of small erratics within the rhythmites (Rigby photo).

Close-up view of two rhythmites, showing varved annual layers of silt and clay within each rhythmite, and unconformity caused by flood scour between the rhythmites (Rigby photo).

Panoramic view of the east side of Latah (Hangman) Creek at the Truck Gardens exposure 3 miles south of Spokane, showing 20 rhythmites (Rigby photos).

CHAPTER 16

THE CHANNELED SCABLANDS

GRAND COULEE AND MOSES COULEE

(Washington 17 and 155, U.S. 2)

Grand Coulee, the most spectacular of all the features produced by
the Bretz floods, was first described in detail by Bretz in 1932. It lies
in the northwest part of the Columbia Plateau, occupying over 250
square miles, enclosed by breathtaking cliffs and desolate
scablands.

The Grand Coulee is a 50-mile-long trench from 1 to 6
miles wide, with steep walls of basalt up to 900 feet high. Its
chiseled course extends south from the Columbia River, just above
Grand Coulee Dam, to the Quincy Basin north of Ephrata. Before
the construction of the dam, the head of this canyon stood 500 feet
above the river, at 1550 feet elevation. Near the middle of this
awesome gorge (U.S. 20) a broad stretch of impressive dry falls
breaks its course, where a vertical escarpment up to 350 feet high
extends for nearly 15 miles east and west, 5 times the width of
Niagara Falls. The greatest and westermost cataract, over 3 miles
wide, was at Dry Falls. Along the cliffs to the east, the visitor can see
5 other reentrant or recessed falls similar to those at American and
Canadian Falls at Niagara. During the Ice Age floods, water over
200 feet deep poured across the lips of these falls. The "plunge
pools" beneath the cliffs, formed by water tearing out the rock, are
now salty kolk lakes.

Here we must diverge for a moment to discuss some new

Aerial view looking northeast to Dry Falls in the upper center, with Banks Lake and the Upper Grand Coulee beyond. Park Lake is on the lower left, and Deep Lake lies below the cataract cliff on the upper right. All of the lower and right half of the picture was under 200 feet of water during flood maxima. The lower part of the picture is occupied by the scabland basalt and the lower cataracts into Jasper Canyon, whose floor is covered with great bars (Delano photo).

ideas concerning plunge pools, pot-holes and any other closed depressions along the course of a stream. Plunge pools do form at the base of waterfalls. But closed depressions may also be formed *beneath* a torrent at the lower end of a steep drop in its course, or where the valley narrows and the volocity if thereby increased. This is how the many small lakes in the coulees of eastern Washington were formed.

The magnitude of the Bretz floods strongly suggests that all these features were formed beneath the surface of the floods, and that there were *no* true waterfalls in eastern Washington during the maximum spate of each flood. The terms "waterfall" and "cataract" that we have been using are in a sense improperly used, since these great cliffs and depressions were entirely beneath the water during the cresting of the floods.

We have no word in our language to use for all these depressions, so, as always, we use a foreign term, this time from the German, "kolk" (deep pool, eddy, scour) or from the Dutch "colk" (a hole eroded by the rushing water at the base of a broken dike) (See Gary, et al. 1973). All the lakes on the Plateau, both at the base of cliffs and along the coulees should properly be called "kolk lakes".

There was probably only a slight drop in the water surface at "Dry Falls" during the height of each flood!

Upper Grand Coulee: Above Dry Falls (State 155), Grand Coulee is 800 to 900 feet deep, 1 to 6 miles wide and 25 miles long. Lieutenant T. W. Simons of the U.S. Cavalry, visiting the Coulee over 100 years ago, wrote ". . . we went north through the coulee, its perpendicular walls forming a vista like some grand old ruined roofless hall, down which we traveled hour after hour." Numerous knobs of basalt and granite, mesas and terraces protrude from its nearly level bottom. Basins along its course are filled with deep gravel fill. Many of those features are now covered by water of Banks Lake Reservoir as a part of the giant irrigation project which now makes eastern Washington one of the productive agricultural regions of the world.

Grand Coulee and its related Moses Coulee a few miles to the west were not, however, formed entirely by the Bretz floods. Rather, when the Cordilleran Ice Sheet came down the Okanogan Valley it crossed the Columbia River and spread out to cover over 500 square miles of the Waterville Plateau plateau to the west of the Coulee. The southern terminus of that glacial advance is clearly marked by a string of east-west trending terminal glacial moraines west of Dry Falls (U.S. 2). The ice dam formed a great Lake Columbia, and water from the lake and river was diverted down a structural depression, a down-fold in the basalt, to eventually carve

Aerial view looking northeast along Banks Lake in Upper Grand Coulee. Small lakes on the Waterville Plateau to the left occupy scour channels in basalt and potholes in the recessional moraines of the Okanogan ice lobe which dammed the Columbia River and diverted it through Grand Coulee. Note that Grand Coulee is now a storage reservoir of water pumped into the Coulee from the Columbia River to support the vast irrigation system in Central Washington. Hence the photographs show water in what 30 years ago was a dry coulee (Delano photo).

out much of the Coulee. The level of Lake Columbia rose to 2400 feet elevation, 1100 feet above the surface of the present Franklin Roosevelt Lake. Much later, the catastrophic Bretz floods found this easy course and deepened and widened the channel.

Spillways to the east: Below the narrows above Dry Falls 4 miles north of Coulee City (U.S. 2), the Bretz floods overflowed the east walls of the main Coulee for nearly 15 miles, depositing a great *expansion bar* in the backwater of Hartline Basin east of Coulee City. It eroded the adjacent uplands into an intricate network of channels as far south as High Hill and Pinto Ridge (north of State 28) in its effort to find a way to the sea. Its volume was such that it was forced to divide into two channels, Dry Coulee to the west and Spring Coulee to the east. Spring Coulee now contains Billy Clapp and Brook Lakes and is a fine scabland canyon over 5 miles long, with castle-like buttes, lateral subsidiary canyons and dry cataracts notching it walls.

Lower Grand Coulee: In addition to the 8 small kolk lakes below Dry Falls, the lower 17 miles of the main Coulee between the falls and the Quincy Basin (State 17) is occupied by an almost continuous string of 4 lakes, 2 to 6 miles long: Park, Blue, Lenore and Soap. The largest, Lenore Lake, fills a 6-mile long scoured-out depression across the axis of a fold in the basalt which crosses the Coulee to form High Hill and Pinto Ridge. The western wall above the lake is over 800 feet high.

Moses Coulee: West of Grand Coulee is Moses Coulee, which parallels lower Grand Coulee about 10 miles to the west. It has a broad straight channel 40 miles long, whose walls along its lower course are up to 1000 feet high. Carved out by meltwater from the ice sheet to its north, it continues southwesterly to enter the Columbia River Valley 15 miles southeast of Wenatchee (State 28). Near the middle part of its course it broadens to the east into a wide basin filled with gravel deposits from the glacier to the north. It shallows north of this basin (U.S. 2) until it reaches its northern end, marked by Grimes and Jameson Lakes.

THE CHENEY-PALOUSE SCABLAND

(U.S. 10, 28, 90 and 395, Washington 23, 26, 124, 260)

Almost unbelievable in size, the Cheney-Palouse Scabland sluiceway, wider than the Amazon river except at its mouth, slopes southwest for 90 miles from Spokane to the Snake River and dwarfs even the majestic Grand Coulee in width and length. It is 10 to 25 miles wide, and covers 1500 square miles, as it drops 1300

Aerial view looking north towards Dry Falls (upper center), Deep Lake (upper right) across Jasper Canyon (lower half of picture). Park Lake and the upper end of Blue Lake lie on the left. Numerous small lakes occupy depressions torn out of the basalt. The entire area of the picture, with the exception of the Waterville Plateau in the upper left, above the visible flood-crest line was covered by 200 feet of water during flood maxima. (Delano photo).

feet in elevation along its course. It has a gradient of 14 feet to the mile, which is half again as steep as the average gradient from Lake Missoula to the sea.

Amerindians, fortunate enough to be above the crest of the Bretz floods, would have seen an endless expanse of tossing waves on the raging brown waters, dotted with rolling and bouncing ice floes and bergs. Only a few islands of loess, rising a few feet above the water, would appear in the upper course below Cheney; farther south, the islands become more numerous, and east of Ritzville the floods split to occupy many channels between 30 elongated islands before reaching the Snake River south of the town of Washtucna.

The scouring and plucking of the bedrock floor within the channels produced depressions now occupied by the hundreds of small lakes dotting the area. Among the largest of these (from north to south) are Medical Lakes north of Cheney; a group of nearly 30 south of Cheney, Sprague Lake (U.S. 10) and 10-mile-long Rock Lake 15 miles south of Sprague.

The Palouse River, which originally flowed down Washtucna Coulee, was diverted south through a deep spillway now occupied by the lower 10 miles of the Palouse River, which at Palouse Falls plunges over a 200-foot-high cliff into a narrow gorge six miles north of the Snake River.

Just west of Lake Kahlotus (Washington 260), another spillway from Washtucna to the Snake formed Devils Coulee. A broad, sloping 45-mile-long plain, known as Eureka Flat, is located south of the Snake River west of Lamar (Washington 124). It was covered by flood gravels, as the waters spilled over the divide between the Snake and Walla Walla River Valleys.

INTERMEDIATE CHANNELS BETWEEN GRAND COULEE AND THE CHENEY-PALOUSE SCABLANDS

Not all the water could be accommodated by the main Cheney-Palouse Scabland and Grand Coulee. Besides flowing down the Spokane River to the Columbia Valley, the floods carved out a multitude of shallow anastomosing channels (a network of splitting and rejoining floodways) cut southward and westward in the soft Palouse Formation loess. These channels are almost too complex to be described! They will be briefly mentioned, from north to south and from east to west.

West of the Cheney-Palouse sluiceway, Lake Columbia,

Aerial view of elongated loess (Palouse Formation) "islands" east of Palouse River and north of La Cross (Washington 26). Looking east. (Carson photo).

Aerial view looking west along Cottonwood Creek, 15 miles south of Sprague, in the upper part of the Cheney-Palouse Scabland near Revere (Carson photo).

Aerial view of scour channels and small lake near Revere, in the upper part of the Cheney-Palouse Scabland. Note scabland buttes, remnants of a stripped lava flow. Looking east (Carson photo).

ponded by the Okanogan ice lobe, spilled southwards across four divides for the 40 miles to Crab Creek and stripped an area of nearly 300 square miles, leaving a "lakes plateau" called the Telford-Crab Creek scabland, dotted with over 40 lakes in the 15 miles southwest of U.S. 2 between Davenport and Creston. Five of the channels are crossed by Highway 21 from Wilbur to Odessa.

The headwater end of upper Crab Creek parallels the Cheney-Palouse scabland and is connected to the scabland by three east-west spillways. From north to south these are Coulee Creek, the two forks of Deep Creek (U.S. 2) 15 to 20 miles west of Spokane, and Rock Creek west of Tyler (U.S. 90). Ten miles north of Ritzville the waters flowed west through a wide spillway into upper Crab Creek Coulee to Odessa and on for another 55 miles (Washington 28) into the Quincy Basin at Soap Lake. Duck Creek and Lake Creek Channels join Crab Creek at Odessa 4 miles to the west. Lake Creek is well named, since there are over 15 lakes, some of them as much as 3 miles in length, along its 35-mile course. An unnamed 25-mile-long coulee occupied by Sullivan and 15 other lakes, joins upper Crab Creek 10 miles west of Lake Creek. Seven miles still farther west is Cannawai Creek Channel.
farther west is Cannawai Creek Channel.

Wilson Creek enters upper Crab Creek just north of Washington 28, two miles west of Cannawai Creek. Crab Creek is the longest and westernmost of the intermediate channels; it extends from its head, only 3 miles south of the Columbia River, for 45 miles across the plateau through Goose Creek at Wilbur on U.S. 2. A second channel, Sinking Creek from the "lakes plateau", joins it just west of Wilbur. One narrow channel flowing from it spills over into the Hartline Basin near Almira.

CHANNELS SOUTH OF UPPER CRAB CREEK

In addition to upper Crab Creek, 6 other west-trending channels were cut across the Plateau from the Cheney-Palouse Scablands to the Quincy and Pasco Basins. Rocky, Farrier, and Bauer Coulees, join to form Weber Coulee, Lind Coulee (U.S. 90) and Washtucna Coulee (Washington 260). Rattlesnake Flat, between Lind and Washtucna Coulees, has been incised on the west by the south-trending Providence Coulee, now occupied by the Burlington Northern tracks between the towns of Lind and Connell. The coulee continues south of Connell as Esquatazel Coulee for another 25 miles to the Pasco Basin.

THE SNAKE RIVER CANYON

Floodwaters surging up the Snake River Canyon helped to erode the late lava flows which had partially filled it, and deposited great mile-long gravel bars high up on its walls many miles upstream from the Palouse River junction. One bar a few miles south of Lewiston, over 100 miles from the mouth of the Snake, is exposed in a quarry, and shows the Bretz flood sands overlying coarse gravels previously deposited by the flood which came down the Snake River from Lake Bonneville in Utah.

This "Bonneville Flood" occurred about 13,000 years ago, when the great Lake Bonneville, which covered much of Utah (Great Salt Lake is the miniscule remnant) broke through a gap at Red River Pass and reduced the level of the lake from 5235 feet to 5135 feet elevation, lowering the lake 100 feet to the Provo shoreline. It has been estimated that the flood released 380 cubic miles of water (about the same as one of the Bretz floods) at a rate of about a third of a cubic mile of water per hour. The foreset bedding in the coarse gravels of the lower part of the bar south of

Lewiston dips downstream, while the finer sand rhythmites in the upper part of the bar dips upstream, showing that the Bonneville flood travelled downstream, and the Bretz floods travelled upstream.

GIANT RIPPLE RIDGES ON BARS

Throughout the scablands, many of the high-level gravel bars have giant ripple-type ridges on their surface, from 10 to 30 feet high, spaced several hundred feet apart. From the ground they are difficult or impossible to see, but they show up clearly on air photographs. After Pardee had shown them to us at Markel Pass in Montana, we knew what to look for, and discovered them on numerous high bars along the Palouse and Columbia Rivers.

CHAPTER 17

LAKE LEWIS BASINS

INTRODUCTION

During the initial surges of the first Bretz flood, torrents made up of a murky, turbulent, ice- and tree-laden slurry rushed across eastern Washington in such great volume that it was unable to escape all-at-once downstream through the Wallula Gap. Up to 264 cubic miles of water were ponded during each of the later, cleaner floods for perhaps a week or so to form Lake Lewis. These temporary lakes covered over 3000 square miles of the Pasco and Quincy Basins; flooded the Yakima and Walla Walla River valleys; and went a hundred miles up the Snake River. These at first coarse and later fine-grained, layered, flood-borne sediments settled out in the relatively quiet waters in these side valleys to form what are called the *Touchet Beds*. Each of the many Bretz floods deposited a layer in these valleys.

WALLA WALLA VALLEY

The Walla Walla Valley is a nearly closed basin, with the river and several more or less parallel tributaries meandering for 30 miles along a broad valley floor to where the river flows through a narrows 8 miles above its junction with the Columbia River. The river and its tributaries rise in elevation from 150 feet below the narrows to 1000 feet at Walla Walla and 1400 feet 6 miles south and east of town. The narrows which restricts the west end of the valley is cut in a wide bench whose surface stands at 600 feet in elevation above the canyon.

In the basin between the tributaries and around the edges of the main valley are several long, narrow, gently west-sloping mesa-like benches 50 to 100 feet above the valley floors, which are collectively called the Gardenia Terrace. The Walla Walla airport lies upon the easternmost of these benches at an elevation of 1200 feet. West of Walla Walla the benches are underlain in part by the fine-grained layered sediments known as the *Touchet Beds,* deposited by the Bretz floods as it dropped its load of debris in this backwater.

Three miles south of Loudon (U.S. 12) an irrigation ditch broke many years ago. It washed out and excavated a narrow ravine widely known as "Burlingame Gulch". This gulch exposes 40 of the 62 layers (herafter called *"rhythmites"*) described by Waitt and Bjornstadt in 1980. Each rhythmite is believed by most geologists to represent a separate flood. Lake Lewis filled an area of the Walla Walla Valley 24 miles long and up to 18 miles wide, covering over 300 square miles. The present site of the town of Walla Walla must have repeatedly been beneath more than 50 feet of water. The waters drained out after each flood through the narrows west of Touchet slowly enough to permit settling of the fine sand and silt into one rhythmite doublet.

A significant thin double layer of white volcanic ash between the 28th and 29th rhythmites in the Touchet Beds at Burlingame Gulch has been firmly identified as the "Set S" ash from Mt. St. Helens. This has been dated at 13,000 years before the present (Mullineaux et al., 1975, 1978). According the Waitt (1983, 1984) the intervals between the forty flood fillings of Lake Lewis averaged about once every 55 years for over 2200 the years between 15,000 and 12,800 years ago.

A similar series of up to 40 rhythmites were discovered in 1965 by Glenn along the banks of the Willamette River on the "Big Bend" between Dayton and St. Paul. They constitute the only evidence so far that *multiple* floods covered the Willamette Valley in Oregon.

Looking southwest across Burlingame Gulch in the Walla Walla Valley, showing flat surfaces of the Gardenia Terrace. The Gulch was cut in the terrace by water from a broken irrigation ditch (Allen photo).

Upper part of the section of 40 rhythmites exposed in Burlingame Gulch. The Mt. St. Helens double ash layers (age 13,000 years) lie between the 29th and 30th rhythmite near the top of the picture. (Allen photo).

Lower part of the Touchet Beds exposed in the Walla Walla Valley beneath the Gardenia Terrace. The earliest floods carried the most sediment (Carson photo).

Granite erratic 10 inches in diameter in place near the base of the section of Touchet Beds in Burlingame Gulch (Allen photo).

Two large gneiss erratics found south of Umapine (Allison photo).

Ten-foot erratic found 5 miles northwest of Umapine, near the State Line (Allison photo).

YAKIMA VALLEY

(U.S. 82 or 97, and Washington 22)

The Yakima Valley extends for over 80 miles to the west and north of Richland, widening westwards from the Chandler Narrows and then narrowing again just south of the town of Yakima. It lies in a broad fold (structural downwarp) in the basalts which forms so much of the underlying rock of Eastern Washington; it is bounded on the south by the Horse Heaven Hills and on the north by Ahtanum Ridge and the Rattlesnake Hills. South of Toppenish it is partially interrupted by a narrows produced by Toppenish Ridge.

The city of Yakima lies north of the main valley in the much smaller Ahtanum Valley, which is linked to the main valley by Union Gap, a narrow canyon cut through Ahtanum Ridge.

Lake Lewis, part of the same lake which inundated the Walla Walla Valley, repeatedly occupied both the Yakima and Ahtanum valleys, rising to an elevation of 1200 feet and covering an area of nearly 600 square miles, with water 200 feet deep above the present city of Yakima. Touchet-like rhythmite beds are widely distributed along the valley floors (Bretz, 1930, 1969, Waitt, 1980) forming a conspicuous terrace 20 to 90 feet high as far north as Parker, just south of Union Gap. The surface of these terraces slopes gently south and east from Parker, where they lie above 800 feet, to an elevation of 300 feet lower near the Chandler Narrows.

Rhythmite sequences, although nowhere as well-exposed as in Burlingame Gulch, occur 1 mile north of Mabton (7 miles south of Sunnyside) and near Zillah, 5 miles northwest of Sunnyside. Thick basaltic gravel deposits at Prosser change progressively upstream to pebble gravel and then to coarse sand at Mabton and are absent at Zillah. Foreset bedding in both gravels and rhythmites dips upstream indicating, as do the similar gravel bars in the Snake River Valley, the enormous volume of water escaping from the main floods to the quieter backwaters of adjoining valleys. Ice-rafted erratics up to 3 feet in diameter are scattered throughout the Touchet Beds in both the Yakima and Walla Walla basins.

QUINCY AND PASCO BASINS

(U.S. 90, Washington 17, 24, 26, 282, 283)

During each flood, Lake Lewis filled the Quincy and Pasco Basins as well as the Yakima and Walla Walla Valleys to elevations approaching 1200 feet, before draining out through the Wallula Gap. Lake Lewis at its maximum extended for 120 miles east and west, and for 60 miles north and south, covering an area of over 2000 square miles. The portion of the lake in the Quincy and Pasco Basins was bounded on the west by the Columbia River Valley wall and Yakima Ridge, and on the south by the Rattlesnake and Horse Heaven Hills. Only the two long east-west ridges of basalt in the Frenchman Hills and Saddle Mountain extend eastward for 30 miles out into the basins. The present sites of Richland and Pasco were submerged under 800 feet of water.

North of Frenchman Hills, the Quincy Basin was filled by water in part from the Grand Coulee and in part from several channels from the east. About half the water came from the north and about half from upper Crab Creek and the channels south of it. Part of the water drained westward from the Quincy Basin into the Columbia River through three channels. These channels can now be seen as the spectacular dry falls at Crater Coulee 5 miles west of Quincy (Washington 28), at the Potholes and at Frenchman Springs coulees 7 and 15 miles southwest of Quincy (Washington 28 and U.S. 90-10).

A network of distributary channels in the northern part of the Quincy Basin radiates out from Soap Lake, east of Ephrata (Washington 17, 28) and joins those from upper Crab Creek. Enormous elongated gravel bars within and between the channels stretch southward for 25 miles to Moses Lake (U.S. 90-10, State 17).

South of Potholes Reservoir and Sullivan Dam (State 11G, 17, 170) a vast labyrinth of butte-and-basin scablands covering 50 square miles is known as Drumheller Channels. They were cut down 300 feet into the basalt by the floods. Dozens of cataracts and narrow interlaced channels and elongated basins trending west to southwest characterize the area. These wild, disorganized scablands continue down lower Crab Creek, to the Columbia River at Beverly (State 243). Floodwaters from the Quincy Basin drained south into the Pasco Basin, mostly through the 15-mile-wide Othello Channels, a small scale copy, 8 miles south of Othello (State 26), of the Drumheller Channels.

The deeper parts of the Quincy and Pasco Basins were, before the floods, filled with sedimentary deposits called the

Ellensberg and *Ringold Formations.* The older Ellenberg fill consists of sediments alternating with or interbedded between some of the late lava flows of Yakima Basalt. The later Ringold Formation lies above the Ellensberg. Both formations were in part eroded, and the area was then largely covered with the widespread flood gravel and boulder debris which filled many of the scoured-out depressions and formed gravel hills and ridges which are flood-produced high gravel bars.

All the scabland channels lead eventually to the Pasco Basin (State 24, 26, 240, 243) It is the largest and deepest depression on the Plateau, more than 500 square miles in area. In the central part of the basin several hundred feet of older sediments and flood deposits lie above the Yakima Basalt, which appear at the surface only in Gable Butte and Gable Mountain, 10 and 20 miles east along the trend of the Umptanum fold south of Priest Rapids Dam.

Gravel bars in profusion characterize the area. One bar south of Gable Mountain is 12 miles long; another, the Priest Rapids Bar (Washington 243), covers 20 square miles. Its summit is over 400 feet above the Columbia River. The enormous Cold Creek Bar, by far the largest of the Pasco Basin gravel bars (Washington 240), is now occupied by the Hanford nuclear installations. It extends southeast from the Columbia River below Priest Rapids Dam for nearly 40 miles to Richland and is from 5 to 10 miles wide. Its summit, just south of the river, stands at 800 feet elevation and must have been covered by 400 feet of water during Lake Lewis maxima.

Wallula Gap from near Hat Rock. Floodwaters nearly overtopped the cliffs bordering the gorge (Sargent photo).

CHAPTER 18

WALLULA GAP TO THE JOHN DAY RIVER

(U.S. 84, 395, 730 , and Washington 14)

INTRODUCTION

The rate of flow of the early floods through the Wallula Gap (I)* has been calculated to have been 1.66 cubic miles per hour, or nearly 40 cubic miles per day for at least 10 days (Weis & Newman, 1973). The speed of the flood with its contained debris must have been up to 50 miles per hour. The water crested near the top of the cliffs in the Gap, and on either side there are overflow channels up to 1250 feet in elevation, some of them containing small bars. The present highway (U.S. 395-730) must have been submerged by over 1000 feet of water!

In comparison, this is 60 times the flow and 320 times the volume of the largest historical flood on the Columbia (1948), which was only 0.7 cubic miles a day for 3 days. The cross sectional area of the gap below 1250 feet elevation is about .326 square miles, which also suggests a flow of over 400 million cubic feet per second.

---

*Roman numerals and capital letters refer to the locations of the narrows, spillways and other features along the Columbia River and south of Portland on the index maps, pp. 133, 146, 168, 176, 193.

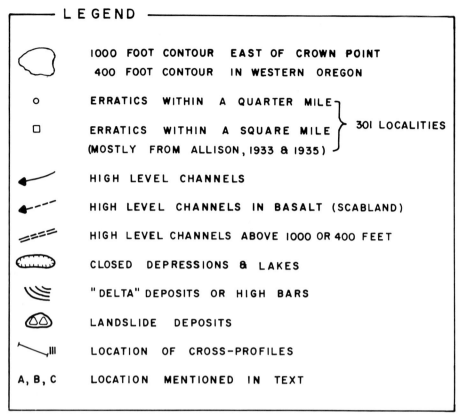

Legend, showing features located on the five maps (pages 133, 146, 168, 176, 193) which cover the extent of the Bretz Floods in Oregon and adjacent Washington.

West cliffs of Wallula Gap, with small tributary spillover channels in center and on left skyline at elevation of about 950 feet (Sargent photo).

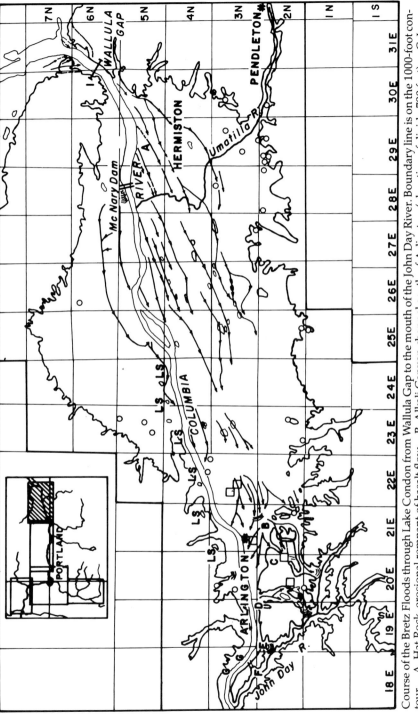

Course of the Bretz Floods through Lake Condon from Wallula Gap to the mouth of the John Day River. Boundary line is on the 1000-foot contour.   A. Hat Rock, erosional remnant of basalt flow.   B. Alkali Canyon channel south of Arlington (elevation of divide 790 feet).   C. Jones Canyon channel (1004 feet).   D. Blalock Canyon channel (960 feet).   E. Philippi Canyon channel (860 and 775 feet).   F. Unnamed spillway (1030 feet).   G. Faceted spurs on north side of river.   Measured cross-section across the Columbia River at Wallula Gap (I). There are 26 erratics or groups of erratics shown.

LAKE CONDON

In the Umatilla and The Dalles Basins the water spread out to form *Lake Condon,* which rose to an elevation of slightly over 1000 feet. Soft sediments and the upper basalt flows in the basin on either side of the river were swept away and reworked over an area of nearly 1500 square miles. As Lake Condon rose, the water poured across the high-level wind gaps south and west of Arlington, up Alkali and Philippi Canyons into Rock Creek and the John Day River Valley. Some 26 exotic boulders (called *erratics*) dropped by stranded melting icebergs around the shore of Lake Condon have been identified.

Every 30 to 60 years, a new flood filled the Umatilla and The Dalles Basins, over an area 10 to 30 miles wide and 100 miles long, extending from Wallula Gap to The Dalles. The 1000-foot contour on the maps will give the reader a good idea of the minimum area of Lake Condon. The 1200-foot contour would more exactly outline the shores of the temporary lakes.

After ripping away the basalt near the Gap to leave only such remnants at Hat Rock (A), the floodwaters scoured a broad area of the older gravels and sand in the eastern part of the basin, leaving hundreds of closed depressions across the Hermiston-Boardman plains.

The new 7½ minute topographic quadrangle maps along the Columbia River in eastern Oregon show details not before apparent on the older 15- and 30-minute maps. They show many closed kolk depressions, only a few tens of feet deep, mostly aligned along southwesterly to westerly-trending channelways which can be traced for tens of miles. Some of these are shown on the map. Further geological field work will undoubtedly find more. Many of the channels are intersected at right angles by the remarkably straight north-trending courses of Fourmile, Sixmile, Eightmile, Pine, Willow and Coyote Creeks, suggesting that these shallow creek valleys were cut after the last Lake Condon drained.

A total of 26 erratics, some of large size, have been found around the edge of Lake Condon (Hodge, 1931, Sargent, personal comm.). A large semicircular protrusion on the north valley wall across the river from Arlington was formed by a landslide, which was probably triggered by the saturation of underlying sediment from a late Bretz flood.

Hat Rock (A) 10 miles east of Umatilla, a flood-eroded remnant of a Yakima lava flow which is seen in the cliffs on the skyline (Ore. Dept. of Transp. photo).

Erratics depostied in Lake Condon.
A. B. South of Boardman.
C. Near Pendleton.
D. Southeast of Echo Junction. (Allison photos)

High terrace gravel deposited beneath flood-eroded escarpment just below McNary Dam. Looking northwest (Ore. Dept. of Transp. photo).

SPILLWAYS INTO THE JOHN DAY CANYON

Hodge (1931) recognized more than 50 years ago that floodwaters had overtopped the low divides between the Columbia River and the headwaters of Rock Creek, as well as the divide directly into the John Day Canyon. The floodwater poured up Alkali Canyon, south of Arlington (Oregon 19), and scoured a channel westward (now occupied by the Union Pacific RR branch line) into Rock Creek 6 miles above its junction with the John Day River (B). Farther west, the Floods poured up Jones Canyon (C), Blalock Canyon (D) and Phillipi Canyon, just east of Quinton (E), where it formed several square miles of scabland and left a high-perched expansion bar on the east wall of the John Day Canyon 10 miles from its mouth. A sixth small spillway lies at 1020 feet elevation, 2 miles northwest of Phillipi Canyon (F).

VALLEY-WALL EFFECTS

A series of valley-wall spurs on the north side of the Columbia River below Quinton were cut away as the valley was widened by the floods, leaving triangular faceted spurs (G). A thick interbed of flow-breccia in the Yakima Basalt on the south side of the valley was eroded over an area of nearly a square mile into a weird expanse of pinnacles, now submerged by the waters behind the John Day Dam.

DEPOSITS IN THE TRIBUTARIES

The 1000-foot contour extends up the John Day River to Thirtymile Creek and up the Deschutes River for 45 miles to a point 10 miles south of Maupin. Hodge (1931) reported a large erratic boulder lying near Sherars Bridge, 7 miles north of Maupin.

Aerial view of Alkali Canyon flood channel 8 miles south of Arlington, looking due west. Note landsliding in foreground, due to saturation of soft sediments. This was the course of the Old Oregon Trail. A modern highway and railroad now occupy the canyon (Sargent photo).

Aerial view of Alkali Canyon flood channel, looking northeast from a point just above Diamond Butte, north of Rock Creek and 12 miles southwest of Arlington. Most of the area in the left foreground was overtopped by the floodwaters (Sargent photo).

Three spillways from the Columbia into the John Day River, located behind ridge on the right, 15 miles west of Arlington. The eroded scabland surface ("Quinton topography") of basaltic flow-breccia in the foreground is now covered by waters of The Dalles Dam. Looking southeast towards (C) Jones Canyon, (D) Blalock Canyon, and (E) Philippi Canyon (Ore. Dept. of Transp. photo).

Aerial view of the entrance to the Philippi Canyon spillway (E) from the Columbia River into the John Day canyon, looking southwest (Sargent photo).

Aerial view looking east across the spillway (E) from the Columbia River (upper left) up Philippi Canyon and across the divide into the John Day Valley (lower right). Note the large expansion bar on the wall of the John Day Canyon in the center (Sargent photo).

Aerial view showing a closer view of the scabland surfaces west of Philippi Canyon (E) and the expansion bar deposits in the John Day Canyon (Sargent photo).

Aerial view of a small high-level floodwater spillway (F) at an elevation of 1030 feet, located 2 miles northwest of Philippi Canyon. Note small kolk lake in the center of the picture. The Columbia River is beneath the airplane, the John Day River canyon crosses the upper third of the picture (Sargent photo).

The height of the flood crests can be judged to be at the base of the triangular faceted spurs, on the north side of the Columbia River just west of Celilo (G). Note the large landslide in the upper right, possibly a result of the saturation of sediments by the floodwaters (Ore. Dept. of Transp. photo).

CHAPTER 19

JOHN DAY RIVER
TO THE DALLES

(U.S. 84 and Washington 14)

The Columbia River, four miles below the mouth of the John Day, passes through a narrows (II), now the site of the John Day Dam. The narrows has a cross-sectional area below the 1100-foot contour of about .265 square miles, 18 percent less than that at Wallula Gap. The flood waters flowing through these narrows and draining Lake Condon moved at the rate of up to 40 miles per hour.

VALLEY-WALL EFFECTS

The valley widens below the John Day Dam (A), and here the floodwaters crested above the lower cliffs on the south wall for a distance of up to three miles from the river, shaping the upland surface into a series of longitudinal ridges and depressions as far west as Celilo. Below the mouth of the Deschutes River, a high eddy bar forms a bench 500 feet above the river.

The remarkable scouring effects in the river channel at Celilo are best shown in photographs taken before the area was covered by the waters behind The Dalles Dam. Several channels where the river almost "turned on edge" in deep grooves wrenched out of the basalt, which gave rise to the name "dalles," the French word for "gutter," a highly appropriate term.

A high bench up to a mile wide was cut into the north wall of the canyon (Washington 14) on which Maryhill Museum and the Stonehenge Monument now rest. Several large ice-rafted erratics have been found here above 800 feet elevation, one of them at 860 feet.

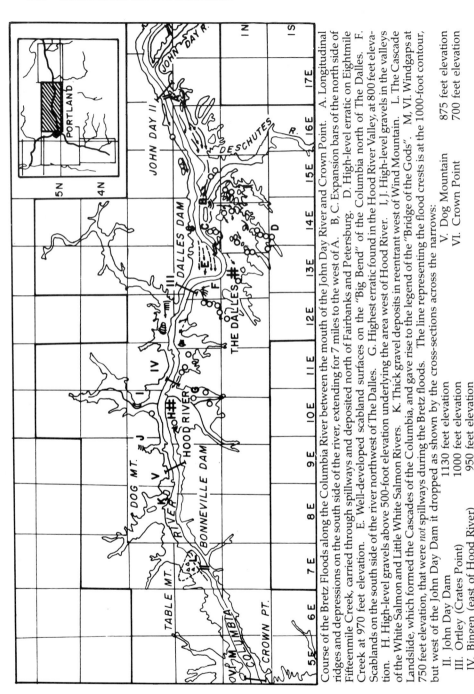

Course of the Bretz Floods along the Columbia River between the mouth of the John Day River and Crown Point.   A. Longitudinal ridges and depressions on the south side of the river, extending for 7 miles to the west of A.   B, C. Expansion bars of the north side of Fifteenmile Creek, carried through spillways and deposited north of Fairbanks and Petersburg.   D. High-level erratic on Eightmile Creek at 970 feet elevation.   E. Well-developed scabland surfaces on the "Big Bend" of the Columbia north of The Dalles.   F. Scablands on the south side of the river northwest of The Dalles.   G. Highest erratic found in the Hood River Valley, at 800 feet elevation.   H. High-level gravels above 500-foot elevation underlying the area west of Hood River.   I, J. High-level gravels in the valleys of the White Salmon and Little White Salmon Rivers.   K. Thick gravel deposits in reentrant west of Wind Mountain.   L. The Cascade Landslide, which formed the Cascades of the Columbia, and gave rise to the legend of the "Bridge of the Gods".   M, VI. Windgaps at 750 feet elevation, that were *not* spillways during the Bretz floods.   The line representing the flood crests is at the 1000-foot contour, but west of the John Day Dam it dropped as shown by the cross-sections across the narrows:

II. John Day Dam   1130 feet elevation   V. Dog Mountain   875 feet elevation
III. Ortley (Crates Point)   1000 feet elevation   VI. Crown Point   700 feet elevation
IV. Bingen (east of Hood River)   950 feet elevation

The location of 32 erratics or groups of erratics is shown.

Two miles west of Maryhill Museum, the Columbia Hills to the north are capped by the small Pleistocene cinder cone, the source of a lava flow which came down the canyon wall and originally reached the river at the east end of Miller Island. On the Maryhill Bench, the material upon which the flow rests is primarily silt from one of the older floods. A person standing on the Oregon shore and looking across the right might believe there were two lava flows from this Maryhill volcano. Actually, it is an illusion. They are two parts of the same flow, joined at the east end. The lower lava scarp is part of a huge landslide block, possibly released by saturation of the underlying silts by the floodwater.

Between Maryhill and The Dalles, prominent strand lines, indicating the position of the highest flood crests, can be seen on the Valley walls on both sides of the river. They can be identified if one knows what to look for. Above the line, the ridge slopes are smooth and rounded, the soil cover is thick. Below the line, rugged outcrops and cliffs of basalt are abundant, and the soil is thin. Above the line, many of the ridge-spurs sloping down from the crest of the canyon wall are cut off abruptly by triangular facets, the base of the triangle resting upon the line. Below the line, the ridge spurs were torn away. In this area the flood-crest line is over 800 feet above the present height of the river.

SPILLWAYS INTO FIFTEENMILE CREEK

At nine and at six miles east of The Dalles, floodwaters 400 feet deep again spilled southwards across the narrow divide between the Columbia River and Fifteenmile Creek through two gaps (B and C), depositing substantial expansion bars whose surfaces are sculptured with giant ripple ridges near Fairbanks and just north of Petersburg (U.S. 197).

Excavations in these bars could produce clues as to the number of floods. Most of them are stratified in coarse sandy or gravelly layers, alternating with fine silty layers, the coarse layers representing the initial surge of each flood and the fine layers the quieter period when water velocities were reduced so that the silt could settle.

During the construction of The Dalles Dam, several perched bars along the river were excavated for sand and gravel. Several of these bars contained crude rhythmite doublets.

Aerial view looking southwest towards Mount Hood from above the mouth of the John Day River, before construction of the John Day Dam. Note the height of the flood-scoured and steepened cliffs and structural benches on both sides of the river (Delano photo).

Aerial view looking west along The Dalles channels and the old Celilo Canal before being drowned by The Dalles Dam, under construction in the upper left (Sargent photo).

Large ice-rafted erratic below highway on the north side of the river at 800 feet elevation near Maryhill (Sargent photo).

Looking northeast across the old Celilo Canal, since flooded by the water impounded by The Dalles Dam. Maryhill Volcano is on the skyline near the center of the picture, with a lava flow coursing down the slope towards the river on the right. Outcrop cliffs on the extreme right have been interpreted as a landslide, dropping down a portion of the lava flow (Ore. Dept. of Transp. photo).

Aerial view showing flood-stripped surfaces of the Priest Rapids flows of the Yakima Basalt on the south side of the river below the mouth of the Deschutes River. The heights of the floods can be judged by the elevation of the uneroded loessal soils on the upland surfaces to the extreme right and left (Sargent photo).

Vertical aerial view of the Columbia River and Fifteenmile Creek, 6 miles east of The Dalles. South of the river on the right is the Fairbanks spillway, showing giant ripple ridges on the expansion bar into Fifteenmile Creek. The mile-wide spillway towards Petersburg lies in the left center, now partially covered with white sand dunes (Soil Cons. Service photo).

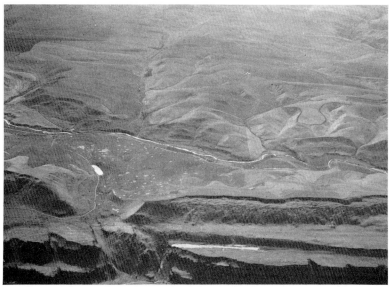

Another aerial view from the north of the Fairbanks spillway into Fifteenmile Creek, 7 miles east of The Dalles. Note the giant ripple ridges above the small kolk lake on the left, and the numerous small depressions on the surface of the broad expansion bar. Extensive landsliding can be seen beneath the upper cliffs in the foreground, as well as along the south side of Fifteenmile Creek (Sargent photo).

Aerial view of the Fairbanks expansion bar, showing the giant ripple ridges in the lower half of the picture. Compare position of the kolk lake in the upper right with its position in the previous photo (Sargent photo).

Aerial view looking northeast across The Dalles Dam. Maryhill volcano lies on the skyline in the upper right, the scablands of the "Big Bend" to the left. Near the right edge of the picture the Petersburg spillway is occupied by an orchard, above it, separated by a ridge which was covered by less than 100 feet of water, lies the Fairbanks spillway (Delano photo).

THE DALLES BASIN

The ponding at the west end of Lake Condon in The Dalles Basin reached the 1000-foot contour. North of The Dalles is the Ortley Gap at Crates Point (III), with a cross-sectional area below this elevation of about .223 square miles, or 15 percent less than the area at the John Day Dam gap. The portion of Lake Condon in The Dalles Basin covered nearly 100 square miles, including the backed-up waters in the valleys of Fifteenmile, Eightmile, Threemile, Mill and Chenowith Creeks. One large erratic (D) is located at 970 feet elevation 10 miles upstream from the mouth of Eightmile Creek (Newcomb, 1970). Deposits of poorly sorted silt, sand, and gravel, with occasional erratics, occur in many of these canyons up to elevations of 800 feet or more. One extensive deposit noted by Hodge (1931) on Chenowith Creek 5 miles west of The Dalles was reported to lie above 1100 feet, but this may be in error.

The Dalles Scablands: The extensive scabland surfaces in The Dalles Basin cover nearly 20 square miles. The "Big Bend" area north of the Columbia River (E) exhibits a multitude of channels, closed depression, small kolk lakes, and low mesa-like ridges between channels. The lowlands beneath The Dalles and for 4 miles down the river (F) show that the floods cut away much of the relatively soft *Dalles Formation* that had once filled the basin, leaving the cliffs at the south edge of town. High-level terraces and the lack of deep soil below 1000 feet elevation north and west of The Dalles can readily be observed as evidence of the flood crests.

Early human occupation in The Dalles Basin:
Since Amerindians were undoubtedly dependent upon fishing for their livelihood, it is likely they occupied the basin before the Bretz Floods and reoccupied the area during the intervals between the floods. Human artifacts have been found in two localities in or beneath flood deposits. The first is in a gravel bar just west of the mouth of the John Day River, from which Luther Cressman recovered a stone knife in 1953. The second locality is near the lower gate of The Dalles Dam navigation lock. There, during the excavation of the lock chamber, a hearth was discovered which yielded several artifacts, charred animal bones, impressions of willow leaves and a small amount of charcoal. The sand in which the hearth rested was overlain by 30 feet of poorly sorted flood gravel. Proof of age is lacking since the sample lacked enough carbon to run a carbon 14 analysis. The bone fragments were later lost during a move of the University of Oregon museum.

Aerial view northwest from above The Dalles. Mount Adams and Mount Rainier can be seen on the upper right. The flood crestline shows up on the slopes of the Ortley anticline (Crates Point) on the left as a change in vegetation and color at the 1000-foot contour. East of the river, faceted spurs and terrace deposits mark this line (Ore. Dept. of Transp. photo).

Flood-crest terraces lying in a reentrant protected by a spur 5 miles north of The Dalles. They are cut in coarse basalt rubble, and contain blocks weighing as much as a few tons (Sargent photo).

CHAPTER 20

THE DALLES TO PORTLAND

(U.S. 84, Oregon 30 and Washington 14)

. . . a wall of water . . . tore away the sides and widened the chasm to its present proportions.

S. C. LANCASTER *(1915)*

INTRODUCTION

It is interesting to note that the widening of the Columbia River Gorge by great floods was recognized 4 years before Bretz by Lancaster, the remarkably prescient engineer who designed the famous Columbia River Scenic Highway, large parts of which are still open to travel (U.S. 30). He may have talked with Bretz, who was in the area on field trips during the construction of the highway.

Through the Columbia River Gorge, the crests of the floods dropped in elevation from 1000 feet at The Dalles to 700 feet at Crown Point. Here it abruptly spread out in the Portland Basin and lowered to 400 feet over the site of the present city. As the torrents rushed through the canyon they tore away the lower slopes and changed the cross-sectional profile across the valley from a "V-shape" to a "U-shape." It cut away the lower parts of the ridges on the south side to form triangular slopes between the tributary valleys, known as "faceted spurs," and left the streams hanging high on the walls to form the waterfalls and cliffs which now contribute so much to its scenic grandeur. The north side of the Gorge was over-steepened and made susceptible to the many landslides that later later devastated as much as 50 square miles of the slopes.

The cross-sectional area below the Dalles at Ortley Gap (III) is .223 square miles, 17 percent less than at the John Day Dam Gap (II) where it is .268 square miles. Through the Gorge the floods flowed at a speed of at least 35 miles per hour, although they must have speeded up again below Crown Point for a few miles.

VALLEY-WALL EFFECTS

Between The Dalles and Hood River, soil and talus were swept from the valley walls, leaving the series of clean steps we now see on the barren lava flows of Yakima Basalt. At many places the accumulations of talus which one would expect to see at the foot of the cliffs have yet to form substantial slopes. Just east of Mayer State Park (U.S. 30) the river makes an abrupt turn to the north. This change in the course of the floods caused the cliffs to be undermined so extensively that a great landslide developed, down which the Scenic Highway swings in a series of loops (Rowena Loops).

The almost vertical 600-foot-high cliffs here were overtopped by at least 200 feet of water, which cleaned off much of the overlying Dalles Formation and left several scabland channels, one of them by the highway containing a small kolk lake. During the last 6000 years, ash eruptions from Mt. St. Helens volcano 70 miles to the northwest covered the barren basalt surfaces to a depth of several feet; subsequently the ash layers have been eroded into a multitude of peculiar low mounds.

West of the downwarp in the basalt at Mosier in the Bingen Gap (IV) we can see clearly the scouring away of the soil up to an elevation of nearly 1000 feet. The cross-sectional area at Bingen Gap below this elevation is .220 square miles, very little less than at the gap north of The Dalles (III), which is .223 square miles.

East of the bridge across the Hood River, a remnant of the old Scenic Highway occupies a half-mile-long high-level channel cut by the floods at an elevation of 350 feet. At the town of Hood River and dowstream, no evidence has been found that the flood crests rose higher than 900 feet in elevation, although the heavier vegetation in the lower Gorge may have obscured such features. The highest erratic found in Hood River Valley (G) lies on a little knob 5 miles south of the river, at an elevation of 850 feet (Newcomb, 1969).

Nine miles west of Hood River at Dog Mountain Gap (V), the narrows below the 875-feet elevation contains a cross-sectional area of .216 square miles, still less than that at Ortley Gap (IV), which is .220 square miles.

Looking west down the Columbia River below The Dalles, through the Ortley gap (III) at Crates Point. The triangular cliff on the left was overtopped by the floods, the high-water mark on the right is shown by the X (Ore. Dept. of Transp. photo).

Aerial view showing oversteepened cliffs in basalt flows north of the Columbia River, 8 miles west of The Dalles, just west of the Ortley fold and water-gap. The flood crest line lies at the top of the highest cliff. Above this elevation (about 1000 feet) the soil has not been stripped off by the floods. A modern highway and railroad now skirt the river. (Sargent photo).

Thick deposits of alternating layers of sand and silt (rhythmites) in a 600-foot elevation eddy bar, west of the mouth of the Klickitat River, located at X on the previous photo. The cross-bedding of the deposits dips upstream, away from the Columbia River (Sargent photo).

Aerial view looking north across the Columbia River towards the mouth of the Klickitat River and Lyle, Washington (on the right). Note the broad unvegetated high-level gravel bars above Lyle, and west of the Klickitat at X.

In the foreground is Mayer State Park, the old Scenic Highway and Rowena Dell. The surface of the mesa (now owned by the Nature Conservancy) was stripped clean by the floodwaters and then covered during the last 6000 years by 3 to 4 feet of Mt. St. Helens ash, which has since been eroded into the peculiar mounds which dot the upland surface of the basalt flow. Note the small kolk lake on the mesa in the left center of the picture (Sargent photo).

Aerial view looking northeast up the Columbia River across Bonneville Dam and the Bridge of the Gods at Cascade Locks. About 1260 A.D. the Cascade Landslide in the lower left shoved the river a mile to the south. The Hood River Valley and Bingen anticline can be seen in the upper right, with the Ortley anticline above it (Delano photo).

GRAVEL DEPOSITS IN THE SIDE VALLEYS

Sands and gravels were deposited by the torrents up to 600 feet in elevation west of the town of Hood River (H). On the north side of the river they are found at similar elevations up the side valleys of the Klickitat, White Salmon (I), Little White Salmon (J) and west of Wind Mountain (K). The cliffs are so steep on the south side of the valley that their talus has probably covered any flood deposits that might have been deposited there.

LANDSLIDES IN THE GORGE

Oversteepening of the valley walls by the Bretz Floods, particularly in the areas on the north side where the Yakima Basalt was underlain by the easily erodable *Eagle Creek Formation,* set the stage for landsliding. Nearly 50 square miles north of the river is characterized by this landslide topography, a hummocky surface with numerous depressions, many of them occupied by lakes (Palmer, 1977). The largest group of slides is the Cascade Landslide north of Bonneville (L), which covers nearly 14 square miles. A large part of this landslide came down about 1260 A.D. (Lawrence, 1958), probably triggered by a great earthquake. It shoved the river south for a mile, creating a temporary dam and lake more than 200 feet deep.

The lake behind such a dam, three times as high as the present Bonneville Dam, might have taken several years to fill before being overtopped by the ponded waters. This temporary dam undoubtedly gave the Indian inhabitants plenty of time to develop the legend of the "Bridge of the Gods." Subsequent erosion by the river of the dam resulted in the "Cascades of the Columbia," first described by Lewis and Clark in 1805, from which the name of the Cascade Range of mountains was derived. The Cascades are now submerged beneath the water behind the Bonneville Dam.

POSSIBLE SPILLWAYS INTO THE
SANDY AND WASHOUGAL VALLEYS

Two possible spillways at elevations of 750 feet into the Washougal River Valley to the north, one between Mount Pleasant and Mount Zion (VI) and the other two miles east of Mount Zion (M), are covered by residual soil and clearly were not overtopped by the Floods, whose crest must have been slightly over 700 feet. A

Historical airview (about 1930) looking north across the great Cascades landslide of 1260 A.D. at Bonneville. The Cascades of the Columbia, Cascade Locks canal and the newly constructed Bridge of the Gods in lower right. The headwall of the landslide can be seen in the face of Table Mountain (left) and Greenleaf Peak (right). Forests had not yet recuperated from the great Yacolt burns of 1910 and 1929 (Brubaker photo from the Oregon Historical Society).

possible spillway at 550 feet elevation into the Sandy River Valley to the south, 3 miles west of Corbett (U.S. 30) also contains a deep soil and was never swept by floodwaters.

From this evidence we can safely conclude the crests of the Bretz Floods, which topped Crown Point (VI) at over 700 feet elevation, dropped steeply to 400 feet as the waters spread out in the Portland-Vancouver Basin. The cross-sectional area below the 700-foot contour at Crown Point (VI) is .215 square miles, still less than that below Hood River at Dog Mountain (V) where it is .220 square miles. Here the waters may have speeded up to perhaps 45 miles an hour.

Aerial view northeast up the Columbia River at Bonneville. Bonneville Dam in the center, the Bridge of the Gods and the town of Cascade Locks are in the upper center. Between them on the left is the lower part of the great 14-square-mile Cascades Landslide, which, about 1260 A.D., shoved the river a mile to the south and produced a 200 foot high dam, which lasted long enough to give rise to the Indian legend of the "Bridge of the Gods" (Delano photo).

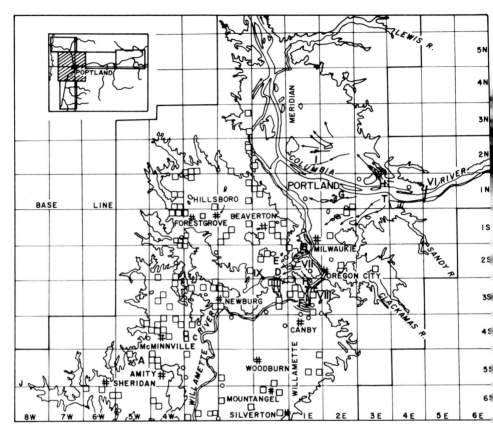

Course of the Bretz Floods (below the 400-foot contour) through the Portland-Vancouver and Tualatin Basins and the northern Willamette Valley. Legend on page 132.

A.　The Belleview erratic, largest ever found in western Oregon. Weight originally about 160 tons, lies at 306 feet elevation.

B.　East end of Lake Oswego flood channel.

C.　Location of "Big Bend" rhythmites.

D.　Tonquin channels and scablands west of Sherwood.

E, F.　Large expansion bars at the west end of the Oswego channel and at the southern end of the Oregon City narrows, north of Canby.

G.　Rocky Butte, with scour channel on east side.

H.　Lackamas Lake floodway north of Camas.

I.　Burnt Bridge Creek channel north of Vancouver.

J.　Divide below 500 feet elevation on the upper Yamhill River.

Measured cross-sections across channelways at:
　VI Crown Point.
　VII Lake Oswego.
　VIII Oregon City.
　IX Tonquin.

Locations of 135 erratics or groups of erratics is shown.

CHAPTER 21

THE PORTLAND-VANCOUVER BASIN

(U.S I-84, I-5, 26, 30, 99-E, 99-W, I-205)

INTRODUCTION

As the floods emerged from the Gorge at Crown Point gap (VI) which is only 2 miles wide, they crested above the Vista House at over 700 feet elevation, then plunged down within about 3 miles to 550 feet elevation, with a width of about 5 miles. The cross-sectional area at Crown Point is .215 square miles, only slightly less than at Dog Mountain (V) of .216 square miles.

Spreading out in the broad Portland-Vancouver Basin, the crests dropped abruptly to about 400 feet elevation, which still left only the tops of Rocky Butte and Mount Tabor above each flood. Again restricted in their flow to the sea by the narrows north of Portland at Kalama (X) they flowed south to fill the Tualatin Valley through the Oswego gap (VII) and the Willamette Valley through the Oregon City gap (VIII) as far south as Eugene, ponding in the valley to form *Lake Allison.*

SUMMARY OF EFFECTS

In the Portland-Vancouver Basin the floods produced the following effects:

1. Below the 400-foot elevation of the flood crests, most of the *Portland Hills Silt* (a deposit of wind-borne loess similar to the *Palouse Formation* of eastern Washington) that had been blown up over the hills during the dry interglacial periods, was almost

Aerial view looking east across the Willamette River and Portland towards Mount Hood. The entire area west of Mount Tabor (left center) was inundated and sculptured by the floods, only this and other prominences like it rose above the waters. One channelway came through south of Kelly Butte (to the right of Mount Tabor) leaving a closed depression which is regularly flooded when Rock Creek overflows. Only the upper 211 feet of the 536-feet-high First Interstate Bank building would rise above the floods if they were to recur today. (Ore. Dept. of Transp. photo).

completely washed away, but still covers the uplands around Portland.

2. Numerous channels were scoured in the underlying Ice Age terrace sands and gravels, leaving kolk depressions such as those around Rocky Butte, north of Kelly Butte and around Powell Butte, as well as north of Vancouver and Camas in Washington.

3. The west- and southwest-flowing torrents over-steepened the upstream sides of Rocky Butte (G), Mount Tabor, Powell Butte, Mount Scott, Kelly Butte and Grant Butte and, west of the river, the lower slopes of the Portland Hills.

4. Throughout the Portland-Vancouver Basin, all the older terraces were covered with a relatively thin skim of flood deposits, which can be seen in the numerous sand and gravel quarries in east Portland. Numerous erratics have also been found in the quarries.

5. Giant ripple-marks were formed along the tops of flood-deposited ridges. The east side of the Reed College Campus, for instance, lies upon one rippled hill 3 miles long and 100 feet high (Palmer, 1982).

NORTH OF THE COLUMBIA RIVER

From Crown Point Gap, the water poured northwestward across the present sites of Washougal and Camas with a depth of nearly 500 feet. It swept up the Washougal and Little Washougal Valleys and across the divide past Woodburn Hill north of Washougal into the valley of Lackamas Creek, where it was joined by a similar torrent which had surged up Lackamas Creek, scouring out the kolk depression now occupied by Lackamas Lake. Only the upper 100 feet of Woodburn Hill and the upper 200 feet of the higher Prune Hill, west of Camas, stood above the waters. The entire Fourth Plains surface was inundated as far north as Woodland, 35 miles to the northwest, covering over 275 square miles of the Basin in Washington.

Northwest of Lackamas Lake (H), one broad channel paralleled the main Columbia River channel courses westward along Burtbridge Creek (I); others split off to the northwest and continued on into Salmon Creek and to the north into the well-developed ridge-and-swale area along Mill Creek.

SOUTH OF THE COLUMBIA RIVER

On the east side of Rocky Butte (G), a deep moat (now occupied by U.S. 85 and 205) was scoured out, much as when a wave retreats around a pebble on the beach, or as a flood undermines a bridge pier. Besides the oversteepening of the east side of the Butte, it cut a vertical cliff in a massive lava flow on the north side. On the lee side of the Butte, two pendant bars trail off westward, from the north side for 5 miles to form the Alameda Crest, and from the south side for nearly 2 miles, leaving a closed depression in the backwater area between them. The gravels and sands deposited in this backwater area between the pendant bars were formerly exposed in a gravel pit east of 82nd Street. Before the pit was filled, the foreset bedding could be seen dipping south from the north bar and north from the south bar.

Another channel known as Sullivan's Gulch, now occupied by U.S. 84, extended west-southwestward from near 57th Street to the Willamette River near the Hawthorne Bridge. To the south of Rocky Butte, Mount Tabor was submerged save for the top 250 feet, and its slopes were steepened on the east side. A channel bit into the north side of Kelly Butte to form a kolk depression at its west end.

A major channelway, running west-southwest from Troutdale and Wood Village, eroded the north side of Powell Butte and flowed into Johnson Creek across a large kolk depression formed at the west end of the Butte between 128th and 136th Streets. The drainage into this depression still creates floods along Johnson Creek every few years. The narrow valley of upper Johnson Creek south and east of Powell Butte was swept by waters nearly 200 feet deep. Gravels deposited alongside and in these channels are still mined in numerous pits.

Between the towns of Milwaukie and Clackamas, a 2-mile wide and 5-mile long northwest-trending double valley was carved out. This channel is now occupied by Kellogg and Mount Scott Creeks and U.S. 224. The low dividing bar between the two parallel creek channels is followed along much of its course by Lake Road. This broad valley must have been occupied by the Clackamas River during much of the ice age. The great mile-long sand and silt bars and many small kolk depressions formerly occupied by lakes are mute testimony not only to great floods but also to the heavy loads of debris the water still carried. It may well be that the northwest end of the valley was filled during the Bretz floods by sediment swept up the ancestral Clackamas River, which diverted it to its present course south of Gladstone.

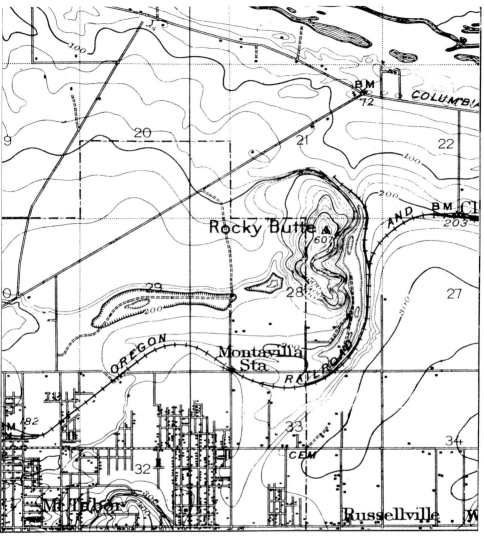

Rocky Butte volcano in east Portland, from a portion of a U.S. Geological Survey topographic map surveyed in 1896. The deep scour channel on the east side (now occupied by U.S. 205) are well shown by hachured contours, as the 1½-mile long closed depression between the two pendant bars west of the Butte. Contour interval is 25 feet.

GIANT MEANDERS IN THE PORTLAND BASIN

One of the rules governing meandering streams is that the size of the meander is related to the volume of water carried by the stream. The larger the stream, the greater the radius of the curve of its meanders. Palmer (1982) has pointed out that in the Portland-Vancouver Basin, the great sweeping curves of the Columbia River have many times the radii of a normal stream of its present size, and that the floodways on the terraces in the south part of the Portland-Vancouver Basin have meanders 2 to 3 times larger than those in the present Willamette Valley. He suggests that this is another line of evidence substantiating catastrophic flooding of the area.

GLACIAL ERRATICS

Numerous erratics up to 5 feet in diameter have been found during the excavation of the gravel pits which dot the Portland-Vancouver Basin. A few are particularly worthy of note.

A 3-foot boulder of gneiss from the Riverdale schoolyard north of Lake Oswego (B) is now on exhibit just east of the Oregon Museum of Science and Industry in Portland. Another is at the rest stop on U.S. 20 just south of West Lynn. Several others were found by one of the writers many years ago in the former county gravel pits just north of Boones Ferry interchange on U.S. 5.

The Mannings kyanite: A unique and significant cobble (6 inches in diameter) was found many years ago in the excavation on the southeast corner of SW 10th. and Morrison Streets in downtown Portland. It is composed entirely of large bladed blue crystals of the rare mineral kyanite. Richmond (1935) reported that "massive slide boulders of kyanite have been found near Revelstoke, B.C. . . ." This is very convincing evidence of its origin in the Purcell Trench of British Columbia; the cobble was surely floated down in one of the Bretz floods, frozen in an iceberg.

The Gladstone baculite: A group of several erratics were found on the top of a low hill at 308 feet elevation a mile northeast of Gladstone. One of them consists of a well-preserved six-inch section of *Baculites,* an extinct shellfish related to the ammonites, found in Cretaceous rocks. It, too, can only have come from rocks of that age in the intermontane trenches of British Columbia.

CHAPTER 22

THE WILLAMETTE VALLEY SOUTH OF PORTLAND

(U.S. I-5, 99-W, 99-E and Oregon 18, 22, 34, 36, 228 and 126)

INTRODUCTION

Each flood filled the Willamette Valley as far south as Eugene to nearly 400 feet in elevation, leaving deposits in the valley which consist of layered silt known as *Willamette Silt* and more than 135 known groups of erratic boulders dropped by melting icebergs around the margin of ephemeral *Lake Allison*. The floods also poured west up the Lake Oswego channel (then the course of the Tualatin River) and diverted the river to its present course. The water filled the Tualatin Valley to at least 300 feet elevation, and spilled over the southern divide between Tualatin and Sherwood through numerous channels which are now the best examples of scablands to be found in western Oregon.

WHY WAS THE WILLAMETTE VALLEY FLOODED?

One not yet fully resolved problem associated with the Bretz floods is the adequacy and nature of the restriction in the Columbia River Valley below Portland which produced ponding in the Portland Basin for a long enough time for the flood waters to sluice through the Oregon City and Lake Oswego gaps and repeatedly fill the Tualatin and Willamette Valleys to nearly 400 foot elevations.

Allen has drawn cross-sectional profiles across the river at all the narrows (I to XI) along its course, and calculated the areas at

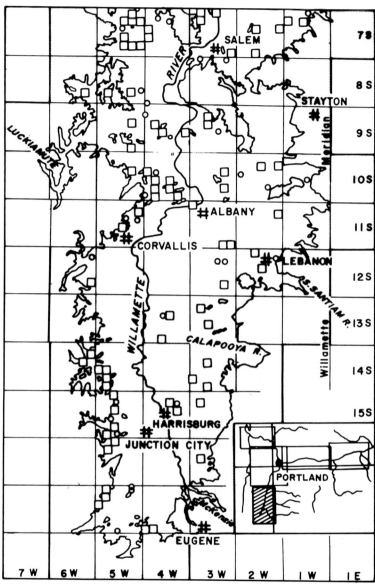

Limits of the Bretz Floods in the southern Willamette Valley (below the 400-foot contour). Legend on page 132. The location of 104 erratics or groups of erratics are mostly taken from Allison (1935).

these gaps lying below the crests of the floods. These calculations have been given throughout PART IV of the text and are summarized in APPENDIX B. Although they generally decrease downstream, as might be expected, they become especially important in solving the problem of Lake Allison.

As an example of illustrating the way the geologic "method of multiple working hypotheses" works, let us discuss the ideas that have been previously proposed to explain this flooding, in the order of their increasing probability.

1. **Landslide dam:** The Columbia River in the Gorge has clearly been dammed in the past by landslides, for example the 200-foot dam which formed the Bridge of the Gods. Such slides are the result of the oversteepening of the valley walls and by saturation of underlying sediments. In addition, lava flows have also dammed the river. Below Portland, however, the walls of the valley do not show evidence of a landslide massive enough to have formed even a low dam, and are certainly not high enough to form a 400-foot landslide dam.

2. **Sedimentary dam:** Glacial outwash, carried down the Cowlitz River drainages from the Mt. Rainier region, has filled large areas south of Chehalis to elevations of over 400 feet at Jackson and Alpha Prairies (A). If this fill once extended down the Cowlitz to Longview, it might have built up a dam there. No evidence of such a fill has been found.

3. **River diversion:** The divide between the Chehalis and Cowlitz River drainage basins (B) is only a few feet above 400 feet in elevation. More than 40 years ago Treasher (1938) suggested that the pre-Pleistocene course of the Columbia River might have been northward into Puget Sound, and that the advances of the Puget lobe of the Cordilleran ice sheet down and across Puget Sound and the Chehalis Valley might have diverted it westward across a low divide in the Coast Range to its present course west of Longview. Such a diversion could scarcely have occurred during the later Bretz floods.

4. **Eustatic (world-wide) rise in sea level:** During the warm interglacial periods, sea levels stood up to 150 feet higher than today, due to the melting of the ice in Greenland and Antarctica. Evidence for such rises in sea level can be found in the marine terraces along the Oregon coast. Such an elevation of the sea would have slowed the flood's descent. During the period of the Bretz floods, however, sea level was as much as 300 feet lower than at present.

5. **Ice dam:** The sheer tonnage of the icebergs and floes, weighted down by their contained rocks and debris, suggested to Hodge (1931) that the fronts of tumbling bergs and water might well have jammed up in valley narrows to produce temporary dams. Since a mixture of water and ice does not flow easily, such large accumulations in a narrows could block the flow. Allison (1933) also proposed such a series of ice dams along the Columbia River and in the scabland channels as an alternative to catastrophic floods, before Lake Missoula had been recognized as the source. The large size of the gaps north and west of Portland make this improbable.

6. **Hydraulic dam:** Any narrowed passage restricts the rate of water flow and causes the water upstream to be temporarily ponded. The cross-sectional area at the Kalama Gap ($X=.163$ miles^2) is 76 percent of that at Crown Point ($VI=.215$ miles^2), while the combined areas of the Lake Oswego ($VII=.036$) and Oregon City ($VIII=.031$) gaps is 31 percent of the area at Crown Point. This third of the total flow was sufficient to fill the Willamette Valley.

The excess water, supplied in part by the accelerated rate of the flow in the drop from over 700 feet at Crown Point to 400 feet over Portland, was sufficient to produce a hydraulic dam at the Kalama Gap.

Of course, the cross-sectional area is only one of many factors which influence the volume of floodwaters! The change in gradient has been mentioned, but other factors, such as the time span during flooding and the rate of flow, have only been roughly estimated. In an attempt to prove that one third of the flow was sufficient to fill Lake Allison, we approached the problem another way.

From a casual study of topographic maps of the area, we can infer the percentage of present land surface covered by the 3000 square miles of Lake Allison below various elevations (column 1). From this, we calculate the percent of the area at each elevation (column 2). The depth of the water determined in percent of a mile is shown in column 3. Multiplying the area by each depth, we can obtain the number of cubic miles of water necessary to fill Lake Allison (column 4).

1 Average elevation in feet	2 Percent of area	3 Area in sq. m.	Depth	4 Percent of mile	Cubic miles
Less than 100	10	300	400	.075	22.50
100–200	25	750	300	.057	56.25
200–300	55	1650	200	.038	62.7
300–400	10	300	100	.019	5.7
Totals		3000			147.15

One third of the volume of 480 cubic miles of water released by a flood is 160 cubic miles. The very rough estimate just made seems to be a reasonable approximation!

This elementary analysis suggests that the last hypothesis, that of the "hydraulic dam," seems to be perfectly feasible. Actually it is the one preferred by most geologists. The inconceivably great volumes of water that repeatedly came down the Columbia River during the last 2000 years of the Ice Age were too great, during the first surge of each flood, to flow down the sea through the Kalama Gap (X). As a consequence, the waters started to back up and in so doing found a convenient outlet by flowing up the Willamette Valley.

As flooding increased to form Lake Allison, some of the icebergs containing rocks (erratics) were carried into this broad lake. These floating bergs with their load of rocks provide much of the evidence to reconstruct the history of the Bretz floods in the Willamette and Tualatin Valleys.

EARLY INTIMATIONS OF FLOODING

The past submergence of the Willamette Valley was so obvious to a geologist that Thomas Condon, the father of Oregon geology, recognized it more than 100 years ago (Condon, 1871) at a time when even the ice age fluctuations of sea level, due to the piling of the ice caps on land and their repeated melting, was still in question. He did not recognize catastrophic or repeated flooding, but assumed a slow rise in sea level to a maximum of 330 feet above present drainage levels. A chapter in Condon's book, *The Two Islands* (1902), the first ever written on Oregon geology is entitled "The Willamette Sound."

GLACIAL ERRATICS IN THE VALLEY

Sixty-four years after Condon, Ira Allison (1935) published an account of those fascinating exotic boulders that have puzzled farmers for over 100 years—one of the most conclusive evidences for the Bretz floods.

> "Many different kinds of rock are represented among the erratics. They include several varieties of coarse-grained igneous rocks, ranging mostly from granite to granodiorite, several varieties of quartzite (white, pink, red, purplish; coarse and fine), slates, phyllites, schists, gneisses, greenstones, rhyolites, pegmatites, porphyies, basalt, diabase, vein quartz, crystalline limestone, and perhaps others. Granite, granodiorite, quartzite, gneiss, slate and basalt are most common. Wherever several pieces occur together, several kinds of rock generally may be found among them."

The erratics range in size from small fragments and chips up to boulders, some of which weigh many tons. They lie scattered over the fields, frequently in clusters or, more commonly, perched on hillsides or on crests of ridges. Few are found below an elevation of 100 feet. There used to be one small group of six boulders on Judkin's Point east of Eugene at 650 feet elevation. Their presence there has been attributed to Indian transport.

Many of the erratics bear scratches (striae), or grooves and faceted faces, an indication that while frozen in ice they had been ground against the bedrock as the glacier moved. Few of the rock types are found in Oregon. To find similar kinds of rock one must go to the Rocky Mountains in Montana and southern Canada. By 1935, Allison had noted over 300 localities where these erratics occur, in 249 different sections (square miles) of land. Because of this early work by Allison, we are here calling the succession of lakes in which the erratics were floated into place frozen in icebergs, "*Lake Allison.*"

The hundreds of icebergs containing the erratics mapped by Allison as far south of Eugene must have been carried in the first surge of each flood. Otherwise they would have had to float the hundred miles up the valley over a period of many weeks and would have melted before reaching their destination. Each lake probably did not last more than a few days or weeks.

Probably less than 50 of the many hundreds of boulders observed by Allison in the 1930's can be found today. They have been broken up for road metal, built into foundations, made into steps, used in rock gardens, or hauled away and buried.

Four large erratics, located about 1 mile south of Wards Butte, 2 miles west of Brownsville (Allison photo).

Four-foot diameter granite erratic on Sevenmile Lane, near Albany (Allison photo).

The Belleview erratic: The largest erratic now known in the valley (A) is an argillite boulder, lying at 306 feet elevation on the top of a low spur half a mile north of Oregon 18 between McMinnville and Sheridan. It would have taken an iceberg measuring at least 34.3 feet on a side to have floated the 160 ton boulder into place! A sign on the highway points out this erratic, and a small state park has been set aside for it. When it was first measured (Hodge, 1950), it was 21 × 18 × 5 feet and weighed about 160 tons. Human "attrition" during three decades has markedly reduced its size to 18 × 13 × 5 feet, with a weight of about 90 tons! If well-meaning but uninformed tourists continue to take "souvenirs" at this rate, this unique geologic phenomenon will have disappeared by the year 2019.

The Willamette meteorite: The Willamette meteorite (H), is the largest ever found in the United States and the sixth largest in the world (Lange, 1965). This 31,107-pound mass of nickel-iron was found in 1902 on the top of a spur on the east side of the Tualatin Valley 2 miles northwest of the town of West Linn at an elevation of 380 feet. This seems to the writers to be an ideal location for the stranding of a small iceberg which came through the Oregon City Gap (VIII) and was carried north up in the Tualatin Valley backwater. In 1985, Richard Pugh discovered several granite erratics only a few feet from the site of the meteorite! The meteorite did not fall in Oregon but originally descended onto the ice lobe in the Purcell Trench or farther north on the Cordilleran ice sheet in Canada. It moved slowly south down the trench to near Lake Pend Oreille and there was carried down in an iceberg during one of the Bretz floods.

RHYTHMITES IN THE WILLAMETTE VALLEY

The *Willamette Silt* first described by Allison (1953) is a widespread deposit of bedded sand and silt which overlies several of the younger land surfaces above the flood plains of the present streams (Parsons, et al. 1969, 1970) in the Valley. It varies in thickness up to possibly 100 feet and in places contains up to 40 rhythmites, like those found in the Walla Walla and Yakima Valleys of Washington. Glenn (1965) first described a section on the east side of the river (C) at the "Big Bend" between Dayton and St. Paul (Oregon 221 or 219). He pointed out that the sediments were mineralogically similar to those of the Touchet Beds of Washington and unlike any locally derived sediments. The lower parts of the Willamette Silt were probably laid down by the earlier Missoula Floods, the culminating Bretz Floods depositing the rhythmites.

The Belleview erratic, the largest ice-borne boulder yet found in the Willamette Valley. Located north of Oregon 18, halfway between McMinnville and Sheridan, it lies at an elevation of 306 feet. It is composed of the metamorphic rock argillite. When it was measured by Hodge (1950), it weighed about 160 tons, "tourist attrition" for 30 years has carried off 70 tons! (Ore. Dept. of Transp. photo).

Roadside sign on Highway 18 describing the Belleview erratic, located on the crest of a ridge about 1500 feet to the north (Ore. Dept. of Transp. photo).

Peaty deposits 20 feet below the surface of "Lake Labish," just north of Salem (I-5), yielded a radiocarbon date of 11,000 years. Similar deposits 16 feet below the surface at Onion Flat, 3 miles west of Tualatin (Oregon 99-W), have been dated at a little over 12,000 years (Glenn, 1965). These lakes were both left after the floods, as a result of damming by flood deposits.

LAKE OSWEGO AND THE TONQUIN SCABLANDS

Before the Ice Age, the Tualatin River joined the Willamette at Lake Oswego (B and VII). The floodwaters from the Portland-Vancouver Basin, restricted by the gap south of Oregon City (VIII), poured westward up the pre-flood valley into the Tualatin Basin and scoured out the kolk basin occupied by Oswego lake, steepened the walls, and formed the scabland pinnacles just north of the lake.

Just west of Lake Oswego they spread out and formed a well-defined expansion bar (E) of coarse, torrentially deposited gravels, containing crude cross-beds dipping to the southwest and west. Granite erratics and pieces of iron ore from the walls of the Lake Oswego Gap could be found in abundance in the gravels while they were being mined for road materials just west of U.S. 5. The area is now covered by industrial developments. The boulder and gravel deposits grade westward into coarse sands at Cipole, north of the town of Tualatin. The southeast edge of the expansion bar is cut by a channel which extends due south down the present Tualatin River Valley. It is obvious that this bar changed the course of the river to the south and also produced the temporary lake which filled to form Onion Flat west of Cipole.

Near Tonquin, between Sherwood and Tualatin, the ponded waters in the Tualatin Basin overtopped a low 2 to 3 mile wide divide (D) in numerous channels, and scoured out the finest example of scablands to be seen in Oregon. The highest of the 12 channelway divides lies at 285 feet, the lowest, in Rock Creek, at 145 feet elevation. Only Rock Creek now flows north; all the rest drain to the south. Within the channels there are 14 scoured-out kolk depressions in an area of 4 square miles. This broad sluiceway continues south for another 5 miles to the Willamette River at Wilsonville (I-5), where another expansion bar was deposited. These channels below the 300 foot contour line have a combined cross-sectional area of .013 square miles (IX), much less than that at Lake Oswego, probably because much of the water could pass through the valley now occupied by the lower Tualatin River, and because of the shallow multiple nature of the Tonquin channels

Aerial view looking east toward the Willamette River and Mount Hood, along the course of the Lake Oswego floodwater channel (VII). Note channels leading off the lower right. These constitute an eastern floodway, extending south for 1½ miles across an expansion bar to the Tualatin River. Note the scabland channel occupied by Iron Mountain Boulevard above the Hunt Club racetract on the center left. All but the upper 90 feet of Iron Mountain lies below the 400-foot contour. Lake Oswego is a fine example of a "kolk lake" (Delano photo).

The Tonquin Scablands, swept by the Bretz floods as they overtopped a low divide between the Tualatin Basin and the Willamette Valley. The 300-foot contour line indicates the height of scouring by the crests of the floods; if submergence rose to 400 feet, it accomplished little erosion. Divides (X) along 12 of the channelways range in elevation from 145 feet on Rock Creek to 285 feet in a channel west of Rock Creek. Kolk lakes and closed depressions are shown by the dark pattern. Within an area of 4 square miles there are 14 of these. The topographic map (contour interval 10 feet) is the northeast part of the Sheridan 7½ minute sheet.

and the consequent slower flow of water.

The combined cross-sectional area of the gaps beneath the 400 foot elevation at Oregon City (VIII) and Lake Oswego (VII) total .067 square miles, or one third of that at the gap at Crown Point (V), which is .215 square miles. These measurements suggest that two thirds of the floods must have gone down the Columbia River to the sea, while one third of the water was sufficient to fill the 3000 square miles of the Willamette Valley as far south as Eugene to an elevation of nearly 400 feet.

There is a low divide (J) in the Coast Range at the headwaters of the South Yamhill River west of Grande Ronde (Oregon 22) with an elevation only a few tens of feet above this level. If Lake Allison had been a little higher, it might have spilled over into the Little Nestucca River!

OREGON CITY TO CANBY

The high cliffs in and south of Oregon City were undermined and steepened by the floods pouring into the gap to the south (VIII). The lowlands on the west side of the river, south of West Linn, were stripped off in a structural terrace now known as the Camassia Preserve. Farther south across the river from Coalca and New Era (Oregon 99-E) an elaborate scabland terraine was developed, first described by Stauffer (1935).

The town of Canby (Oregon 99-E) lies upon one of the largest expansion bars in the state (F), covering over 10 square miles. This and the expansion bar at Wilsonville are the farthest south that have been found to date. Farther to the south, evidence of flooding consists mostly in the presence of the *Willamette Silt* and the erratics already described.

MAMMOTHS IN THE VALLEY

Numerous partial skeletons of the mammoth, which became extinct about 10,000 years ago, have been found in the Willamette Valley, and more are being found every few years. Some of them were found during plowing or in shallow excavations; some were found buried beneath tens of feet of silt. Mammoth must have been as abundant here as the elephant used to be in Africa; there may have been dozens of herds of these great beasts roaming the glades between groves of oak, alder and cottonwood. Is it possible that the Bretz floods contributed to their demise in the valley?

Aerial view looking southeast across the Willamette River at Albany toward the Cascade Range. Prominences which rose above the floodwaters are Ridgeway Butte northeast of Lebanon (upper left), Peterson Butte (upper center), Ward Butte (right) and the Powell Hills west of Brownsville (upper far right) (Delano photo).

CHAPTER 23

PORTLAND TO THE PACIFIC OCEAN

(In Oregon, U.S. 30; in Washington, U.S. I-5 and 830)

The narrows near Kalama (X) restricted the outflow from the Portland-Vancouver Basin at a gap, which, below the 400-foot contour, has a cross-sectional area of .163 square miles, 24 percent less than that at Crown Point (VI), where it is .215 square miles. The cross-sectional area below the 275-foot contour at Clatskanie (XI) 12 miles west of Kalama is .155 square miles, only 5 percent less than that at Kalama.

The table in Appendix B shows that the water must have been 400 feet deep between Portland and the Kalama narrows, 275 feet deep at Clatskanie, and near sea level at Astoria. Woodland and St. Helens were both beneath 400 feet of water. Interpolating between Kalama and Clatskanie suggests that both Kelso and Longview were at least 300 feet below the surface of the floods.

Oversteepened valley-walls and faceted spurs are present along the face of the Portland Hills for the 20 miles between Portland and St. Helens and reappear for 4 miles to the south and from 2 to 6 miles west of Rainier. Scablands underlie St. Helens and Goble on the Oregon side and are prominent around Kalama. Still farther west, Nicolai Ridge (C) opposite Cathlamet, Washington is an outstanding example of a basalt ridge undercut by flood erosion resulting from an abrupt north bend of the course of the river. Abnormal steepening continues for another 5 miles on the south side of the river to Aldrich Point (D).

Below Kalama (X) the gradient of the crest of each flood surge, here assumed to be more than 10 feet per mile, was actually steeper than in the Columbia River Gorge, and the crest level

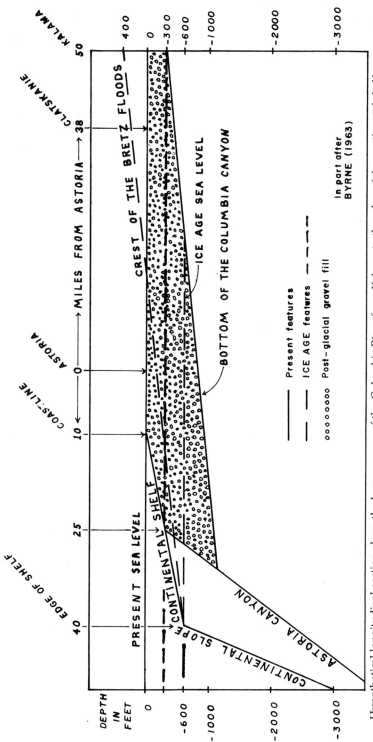

Hypothetical longitudinal section along the lower course of the Columbia River from Kalama to the edge of the continental shelf at present (solid lines) and during the Bretz floods (dashed lines). The deep valley fill has accumulated during the last 12,000 years. If the crests of the floods and the bottom of the valley had slopes of 10 feet per mile as shown, there would be 300 feet of fill at Kalama and nearly 800 feet at Astoria!

dropped rapidly as the valley widened westward to the lowered sea level. Under the assumption that sea level was 300 feet lower than today, the crests of the floods dropped to near present sea level in the 38 miles between Clatskanie and Astoria, while the bedrock bottom of the canyon must have been several hundred feet deeper than that. Continuing westward into the Astoria submarine canyon, the crest of the flood would approach the lowered sea level about 40 miles west of Astoria, near the edge of the continental shelf.

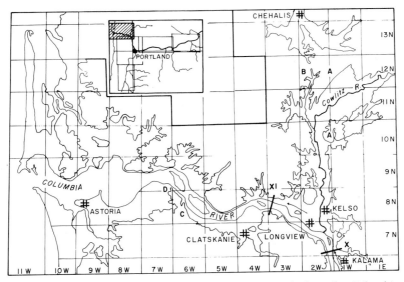

Course of the Bretz Floods (below the 400-foot contour) along the Columbia River between Kalama and the Pacific Ocean (Legend on p. 132).   A. Valley fill of glacial outwash from Mount Rainier above 400-foot elevation.   B. Divide between Cowlitz and Chehalis drainages, below 500 feet elevation.   C. Nicolai Ridge, oversteepened on the east side by floodwaters.   D. Aldrich Point, westernmost extent of known flood effects.   Measured sections across the Columbia River at: X North of Kalama;   XI Northeast of Clatskanie.   Location of one erratic shown.

Looking northeast across the north end of Puget Island from Bradley State Park on Nicolai Ridge (U.S. 30, 25 miles east of Astoria). Beyond the island is the Cathlamet channel, with the town a little to the right of the center of the picture, and Crown Point, Washington (elevation 1815 feet) on the skyline. A north-trending resistant layer of Yakima Basalt diverted the river to the north, and the floods undercut and oversteepened the east side of Nicolai Ridge (Ore. Dept. of Transp. photo).

Appendices

APPENDIX A

AREAS COVERED BY THE EPHEMERAL LAKES

The ponded waters above the Columbia River Gorge in Oregon, Washington, Idaho and Montana covered a total area of over 7700 square miles, divided as follows:

Lake Missoula	2900	*Lake Condon*	
Lake Lewis		Umatilla Basin	1300
Pasco & Quincy Basins	2050	Intermediate and tributary channels	150
Walla Walla Basin	330	The Dalles Basin	100
Lower Snake River	300		1500
		Willamette Valley	3000
		Total	10,710 square miles

The longitudinal profile of the flooded Columbia River from Lake Missoula to the sea (page 105) shows the drop in the flood crests from Lake Missoula (4200 feet) to Lake Lewis (1250 feet) to Lake Condon (1000 feet) to Lake Allison (400 feet) and to the sea at more than 300 feet below present sea level. Half of the total drop took place west of the Gorge, where the canyon widens abruptly into the Portland-Vancouver Basin.

CALCULATIONS OF THE HEIGHT OF THE FLOOD CRESTS

The accompanying table and graph show the calculated (and inferred) data on the progression of the floods from the ice dam to the edge of the continental shelf. The columns display the following:

1. *Distance* from Lake Missoula dam to the edge of the continental shelf.
2. *Locations* of the measured gaps and interpolated points, and the distances between them.
3. *Measured elevations* of the maximum flood crests at the gaps and at interpolated points (inferred), and the vertical distances between them.
4. *Gradients* in feet per mile between all points.
5. *Cross-sectional areas* measured beneath the flood crests at each gap, in square miles.

Each *intermediate* line indicates the *differences* between the stations in miles, elevations of flood crests, and the gradient or slope of the flood crests, in feet per mile. Interpolated elevations are in parentheses.

1	2	3	4	5
0	Lake Missoula dam	4200		
10		700	70.0	
10	Lake Pend Oreille	(3500)		
35		835	24.0	
45	Lake Couer d'Alene	2665		
24		165	6.9	
69	Spokane	(2550)		
60		550	9.0	
129	Ritzville	(2000)		
75		750	10.0	

1	2		3	4	5	
204	Pasco		(1250)			
12			0	0		
216	Wallula Gap		1250		I	.314
25			40	1.6		
241	Umatilla		(1210)			
40			30	.75		
281	Arlington		(1180)			
24			40	1.66		
305	John Day River		1140			
4			10	2.5		
309	John Day Dam gap		1130		II	.268
10			60	6.0		
319	Deschutes River		1070			
13			70	5.2		
332	The Dalles		(1000)			
3			0	0		
335	Ortley gap		1000		III	.223
11			50	4.6		
346	Bingen gap		950		IV	.220
3			25	8.3		
349	Hood River		(925)			
8			50	6.3		
347	Dog Mountain gap		875		V	.216
13			45	3.46		
360	Bonneville		(830)			
15			130	8.66		
375	Crown Point gap		700		VI	.215
7			160	22.8		
382	Troutdale		(540)			
13			140	10.8		
395	Portland		400			
38				0		
	403	Oswego	400		VII	.036
		8 from Portland				
	408	Oregon City	400		VIII	.031
		12 from Portland				
	419	Tonquin	300		IX	.013
		15 from Portland				
	505	Eugene	(380)			
		110 from Portland				
433	Kalama		400		X	.163
12			125	10.4		
445	Clatskanie		275		XI	.155
38			375	10.7		

1	2	3	4	5
483	Astoria	(000)	10.0	
	25	100		
508	Astoria Canyon	(−100)		
	15	200	10.0	
523	Edge of shelf	(−300)		

APPENDIX C

ENERGIES OF CATASTROPHES

ENERGY EXPENDED BY THE FLOODS

Why does the subtitle of this book refer to "cataclysms"? Webster's definitions of "cataclysm'"* are 1) a flood of water, a deluge; 2) any violent change involving sudden and great alterations of the earth's surface.... The Bretz floods affected more than 16,000 square miles of the Northwest. This certainly rates along with the largest earthquakes, volcanic eruptions, great meteor falls and atomic bombs, as causing major changes in the earth's surface.

Calculations of the energy produced by dropping at least 380 cubic miles of ice- and rock-laden water from an elevation of 4200 feet in Montana to 300 feet below present sea level, would amount to 1.9×10^{26} ergs. What does this mean, in terms that we can really comprehend?

It is 26 percent of all the energy produced in the United States during the year 1970. It is equal to 106 percent of all the electrical energy generated from all sources in 1970. It is 77 percent of the energy produced from natural gas, 105 percent of that from oil, and 119 percent of that from coal. It is 7 times all the hydroelectric power and 90 times all the atomic energy generated.

---

*Derived from the Greek: cata=downward; klysein=to wash.

COMPARISONS WITH OTHER CATASTROPHES

	TNT	M	ERGS
One ton of TNT (Ton=T)	1 T		4.18×10^{16}
Earthquake felt (Magnitude=M)	6 T	4.0	2.5×10^{17}
Portland earthquake (1962) largest felt	189 T	5.0	7.9×10^{18}
1000 tons of TNT (Kiloton=KT)	1 KT		4.18×10^{19}
San Fernando earthquake (1971)	1.4 KT	6.2	5.00×10^{20}
Fission bomb (Hiroshima)	20 KT		8.36×10^{20}
Fusion bomb (Bikini)	200 KT		8.4×10^{21}
1 million tons of TNT (Megaton=MT)	1 MT		4.18×10^{22}
Meteor Crater, Arizona	1.7 MT		7.1×10^{22}
Tungska comet (Siberia, 1908)	4.3 MT		1.8×10^{23}
Mt. St. Helens, (1980) 1st 3.3 sec.	5.7 MT		2.4×10^{23}
San Francisco earthquake (1906)	12 MT	8.2	5.0×10^{23}
Alaskan earthquake (1964)	24 MT	8.4	1.0×10^{24}
Largest fusion bomb (USSR test)	60 MT		2.5×10^{24}
Largest recorded earthquakes (Japan, 1933, Columbia, 1906)	134 MT	8.9	5.6×10^{24}
Eruption of Krakatoa (1883)	200 MT		1.0×10^{25}
Mt. St. Helens (1980) 1st 8 hrs.	430 MT		1.8×10^{25}
Eruption of Crater Lake	550 MT		2.3×10^{25}
Bretz flood (largest)	4500 MT		1.9×10^{26}
Eruption of Katmai, Alaska (1912)	4800 MT		2.02×10^{26}
Total energy produced in U.S. (1970)			7.27×10^{26}
Eruption of Tamboro (1815)	20,000 MT		8.4×10^{26}
Eruption of Santorini (1470 B.C.)	24,000 MT		1.0×10^{27}
Meteorite (66 million years ago)	96,000 MT		4.0×10^{27}
Forty Bretz floods (total)	180,000 MT		7.6×10^{27}

In part after Winkless, (1980)

BIBLIOGRAPHY FOR THE LAY READER

Allen, J. E., 1984, The magnificent gateway—a layman's guide to the geology of the Columbia River Gorge: Timber Press, 144 p. (2nd edition)

Baldwin, Ewart M., 1964, Geology of Oregon (Third edition): Kendall/Hunt Publishing Co., Dubuque, Iowa, 170 p.

Bretz, J H., 1928c, The Channeled Scabland of eastern Washington: Geog. Rev., v. 18, p. 446–447.

_____ , 1932, The Grand Coulee: American Geog. Soc. Spec. Pub. 15, 89 p.

Condon, T., 1871, The Willamette sound: Overland Monthly, v. 7, p. 393–398, reprinted in Oregon Geology, 1910.

_____ , 1902, The two islands: J. K. Gill Co., Portland, 211 p.

Gould, S. J., 1970, Is uniformitarianism useful?: (In) Adventures in earth history, W. H. Freeman and Co., p. 51–53.

Hodge, E. T., 1950, The Belleview erratic: Geol. Soc. Oregon Country, v. 16, n. 11, p. 92–94.

Lancaster, S. C., 1915, The Columbia, America's great highway: Private edition, Kilham Stationery and Printing Co., 145 p.

Lawrence, D. B. and E. G., 1958, Bridge of the gods legend, its origin, history and dating: Mazama, v. 11, n. 13, p. 33–41.

McKee, Bates, 1972, Cascadia, the geologic evolution of the Pacific Northwest: McGraw-Hill Book Co., 394 p.

Sargent, S. C., 1953, Tree-rafted erratics along the Columbia River: Geol. Soc. Oregon Country News Letter, v. 19, n. 6, p. 63–64.

Shutler, Richard, Jr., 1983, Early man in the new world: Sage Publications, Beverly Hills, California.

Stauffer, James, 1956, Late Pleistocene flood deposits in the Portland area: Geol. Soc. Oregon Country News Letter, v. 22, n. 3, p. 21–31.

Strong, Emory, 1959, Stone age on the Columbia River: Binfords and Mort, 254 p.

Treasher, R. C., 1938, A Pleistocene damming of the lower Columbia River (abst.): Geol. Soc. Oregon Country Newsletter, v. 4, n. 24, p. 271.

Weis, P. and Newman, W. L., 1971, The channeled scabland of eastern Washington—the geologic story of the Spokane flood: U.S. Geol. Survey pamphlet, 25 p.

Williams, Ira A., 1923, The Columbia River gorge, its geologic history interpreted from the Columbia River highway: Min. Res. of Oregon, Oregon Bur. of Mines and Geol., 130 p.

Winkless, N., and Browning, I., 1980, Climate and the affairs of men: Fraser Publishing Co., 228 p.

TECHNICAL BIBLIOGRAPHY

Alden, W. C., 1927, Discussion; Channeled Scabland and the Spokane flood: Washington Acad. Sci. Jour., v. 17, n. 8, p. 203.

Allen, J. E., 1932, Contributions to the structure, stratigraphy and petrography of the lower Columbia River Gorge: M.A. thesis, Univ. of Oregon, 96 p.

———, 1958, Columbia River Gorge, Portland to the Dalles: (In) Guidebook for Field Trip Excursions, Cord. Sec., Geol. Soc. America, U. of O., p. 4–23.

———, 1984, The magnificent gateway—a layman's guide to the geology of the Columbia River Gorge: Timber Press, 144 p. (Second edition)

Allison, I. S., 1932, Spokane flood south of Portland, Oregon: Geol. Soc. American, Bull. 43, p. 133–134.

———, 1932, New version of the Spokane flood: Geol. Soc. America Bull., v. 44, p. 675–722.

———, 1935, Glacial erratics in the Willamette Valley: Geol. Soc. America Bull., v. 46, p. 605–632.

———, 1941, Flint's fill-hypothesis for channeled scabland: Jour. Geol., v. 49, p. 54–73.

———, 1978, Late Pleistocene sediments and floods in the Willamette Valley: Oregon Dept. Geol. & Min. Ind., Ore Bin, v. 40, nos. 11 & 12, p. 177–202.

Alt, D., and Chambers, R. L., 1970, Repetition of the Spokane flood: American Quaternary Assn., (AMQUA) Abstracts, p. 1.

Alwin, J. A., 1970, Clastic dikes of the Touchet Beds, south eastern Washington: Master's thesis, Washington State University, Pullman, Washington, 87 p.

Anderson, Charles A., 1963, Simplicity in structural geology: (In) The fabric of geology, Freeman Cooper and Company, p. 175–183.

Atwater, Brian F., 1984, Periodic floods from glacial Lake Missoula in the San Poil arm of the glacial Lake Columbia, northeastern Washington: Geology, v. 12, n. 8, p. 464–467.

Baker, H. B., 1938, The fabric of geology: Freeman, Cooper and Company, p. 266.

Baker, V. R., 1973, Paleohydrology and sedimentology of Lake Missoula flooding in eastern Washington: Geol. Soc. American Special Paper 144, 79 p.

_____ , (Ed.) 1981, Catastrophic flooding, the origin of the channeled scabland: Benchmark Papers in Geology, v. 55, Dowden, Hutchinson & Ross, Inc., 360 p.

_____ , and Nummendal, D., 1978, The channeled scabland: Nat. Aeronautics and Space Adm., 186 p.

Baldwin, Ewart M., 1964, Geology of Oregon (Third edition): Kendall/Hunt Publishing Co., Dubuque, Iowa, 170 p.

_____ , 1957, Drainage changes of the Willamette River at Oregon City and Oswego, Oregon: Northwest Science, v. 31, n. 3, p. 109–117.

Bjornstadt, B. W., 1980, Sedimentology and depositional environment of the Tonchet Beds, Walla Walla River Basin, Washington: Master's thesis, Eastern Washington University, 120 p.; also in, RHO-BWI-SA-44, Rockwell International, Richland, Washington.

Bretz, J H., 1913, Glaciation of the Puget Sound region: Wash. Div. Mines and Geology Bull., n. 8, 244 p.

_____ , 1919, The late Pleistocene submergence in the Columbia Valley of Oregon and Washington: Jour. Geology, v. 27, p. 489–505.

_____ , 1923a, Glacial drainage on the Columbia Plateau: Geol. Soc. America Bull., v. 34, p. 573–608.

_____ , 1923b, The Channeled Scabland of the Columbia Plateau: Jour. Geology, v. 31, p. 617–649.

_____ , 1924, The Dalles type of river channel: Jour. Geology, v. 32, p. 139–149.

_____ , 1925, The Spokane flood beyond the Channeled Scabland: Jour. Geology, v. 33, p. 97–115, p. 236–259.

_____ , 1927a, Channeled Scabland and the Spokane flood: Washington Acad. Sci. Jour., v. 17, n. 8, p. 200–211.

_____ , 1927b, The Spokane flood; a reply: Jour. Geology, v. 35, p. 461–468.

_____ , 1928a, Alternative hypothesis for Channeled Scabland: Jour. Geology, v. 36, p. 193–223, p. 312–341.

_____ , 1928b, Bars of the Channeled Scabland: Geol. Soc. America Bull., v. 39, p. 643–702.

_____ , 1928c, The Channeled Scabland of eastern Washington: Geog. Rev., v. 18, p. 446–477.

_____ , 1929, Valley deposits immediately east of the Channeled

Scabland of Washington: Jour. Geology, v. 37, p. 393–427, p. 505–541.

_____ , 1930a, Lake Missoula and the Spokane flood: Geol. Soc. America Bull., v. 41, p. 92–93.

_____ , 1930b, Valley deposits immediately west of the Channeled Scabland: Jour. Geology, v. 38, p. 385–422.

_____ , 1932a, The Grand Coulee: American Geog. Soc. Spec. Pub. 15, 89 p.

_____ , 1932b, The Channeled Scabland: 16th. Internat. Geol. Congress, Guidebook 22, Excursion C-2, 16 p.

_____ , Smith, H. T. U., and Neff, G. E., 1956, Channeled Scabland of Washington; new data and interpretations, Geol. Soc. America Bull., v. 67, p. 957–1049.

_____ , 1969, The Lake Missoula floods and the Channeled Scabland: Jour. Geology, v. 77, p. 505–543.

Bunker, R. C., 1980, Catastrophic flooding in the Badger Coulee area, south-central Washington: M.A. Thesis, University of Texas, Austin, Texas, 184 p.

_____ , 1980, Catastrophic flooding in Badger Coulee, south-central Washington: facies, paleohydraulics and timing: Geol. Soc. America, Abstracts with Programs, v. 12, n. 3, p. 99.

Byrne, John V., 1963, Geomorphology of the continental terrace off the northern coast of Oregon: Ore. Dept. of Geol. and Min. Ind., Ore Bin, v. 25, n. 12, p. 201–209.

Carson, R. J., McKhaun, C. F., and Pizey, M. H., 1978, The Touchet Beds of the Walla Walla Valley: (In) Baker, V. R. and Nummedal, D., 1978, The channeled scabland: Nat. Aeronautics and Space Adm., p. 173–177.

Chamberlin, T. C., 1888, The rock-scorings of the great ice invasions: U.S. Geol. survey, Ann. Report 7, p. 147–248.

_____ , 1897, The method of multiple working hypotheses: Jour. Geol. v. 5, p. 837–843.

Chambers, R. L., 1971, Sedimentation in Lake Missoula: M.S. Thesis, Univ. of Montana, Missoula, Montana, 100 p.

Condon, T., 1871, The Willamette sound: Overland Monthly, v. 7, p. 393–398, reprinted in Oregon Geology, 1910.

_____ , 1902, The two islands: J. K. Gill Co., Portland, 211 p.

Cressman, L. S., 1960, Cultural sequences at The Dalles, Oregon: Trans. Amer. Phil. Soc., v. 50, n. 10, 108 p.

Curry, R. R., 1977, Glacial geology of Flathead Valley, and catastrophic drainage of glacial Lake Missoula. Discussion: Field Guide No. 4, Geol. Soc. America, Rocky Mountain Section Mtg., Dept. Geol., Univ. Montana, p. 31–38.

Davis, W. M., 1926, The value of outrageous geological hypotheses: Science, v. 63, p. 463–468.

Flint, R. F., 1938, Origin of the Cheney-Palouse Scabland tract: Geol. Soc. America Bull., v. 49, p. 461–524.

Gary, M., McAfee, R. Jr., and Wolf, C. L., editors, 1972, Glossary of geology: American Geological Institute, 857 p.

Gilbert, G. K., 1890, Lake Bonneville: U.S. Geol. Survey, Monograph 1, 438 p.

Gilluly, J., 1927, Discussion: Channeled Scabland and the Spokane flood: Washington Acad. Sci. Jour., v. 17, n. 8, p. 203–205.

Glenn, J. L., 1965, Late Quaternary sedimentation and geologic history of the north Willamette Valley, Oregon: Oregon State University, Ph.D. dissertation.

Gould, S. J., 1970, Is uniformitarianism useful?: (In) Adventures in earth history, W. H. Freeman and Co., p. 51–53.

Hanson, L. G., 1970, The origin and development of Moses Coulee and other scabland features on the Waterville Plateau, Washington: Ph.D. dissertation, Univ. of Washington, Seattle, 137 p.

Hobbs, W. H., 1947, The glacial history of the Scabland and Okanogan lobes, Cordilleran continental glacier: Privately printed, J. W. Edwards, Ann Arbor, Michigan, 40 p.

Hodge, E. T., 1931, Exceptional moraine-like deposits in Oregon: Geol. Soc. America Bull., v. 42, p. 985–1010.

_____, 1934, Origin of the Washington scabland: Northwest Science, v. 8, p. 4–11.

_____, 1950, The Belleview erratic: Geol. Soc. Oregon Country, v. 16, n. 11, p. 92–94.

Hutton, J. 1788, Theory of earth.

_____, 1795, Theory of earth with proofs and illustrations: Edinburg, Scotland, 2 vols.

Kroeber, A. L., 1934, Native American population: American Anthropologist, v. 36, p. 24.

Lancaster, S. C., 1915, The Columbia, America's great highway: Private edition, Kilham Stationery and Printing Co., 145 p.

Large, John, 1922, The glaciation of the Cordilleran region: Science, v. 17, n. 1447, p. 335–336.

_____, 1943, Confusion over glacial Lake Spokane: op. cit., v. 98, n. 2556, p. 560–561.

Lawrence, D. B. and E. G., 1958, Bridge of the gods legend, its origin, history and dating: Mazama, v. 11, n. 13, p. 33–41.

Lawson, A. C. (In) 1970, Vaughan, F. E., Andrew C. Lawson, Scientist, teacher, philosopher: Arthur H. Clark Co., Glendale, California, 474 p.

Lenz, R. T., 1977, The petrology and stratigraphy of the Portland Hills Silt: M.S. thesis, Portland State Univ., 144 p.

Lowry, W. D. and Baldwin, E. M., 1952, Late Cenozoic geology of the lower Columbia River valley, Oregon and Washington: Geol. Soc. America, Bull., v. 63, p. 1–26.

Malde, H. E., 1968, The catastrophic late Pleistocene Bonneville flood in the Snake River Plain, Idaho: U.S. Geol. Survey Prof. Paper 596, 52 p.

Macgowan, K., and Hester, J. A. Jr., 1962, Early man in the new world: Anchor Books, Doubleday & Co., 333 p.

McKee, Bates, 1972, Cascadia, the geologic evolution of the Pacific

Northwest: McGraw-Hill Book Co., 394 p.

McKnight, E. T., 1927, The Spokane flood: A discussion: Jour. Geology, v. 35, p. 453–460.

Meinzer, O. E., 1927, Discussion: Channeled scabland and the Spokane flood: Washington Acad. Sci. Jour., v. 17, n. 8, p. 207–208.

Mullineaux, D. R., Hyde, H. J., and Rubin, M., 1975, Widespread late glacial and post-glacial tephra deposits from Mt. St. Helens volcano, Washington: U.S. Geol. Survey Jour. Research, v. 3, n. 3, p. 329–335.

_____, Wilcox, Ray E., Ebaugh, Walter F., Fryxell, Roald and Rubin, Meyer, 1978, Age of last major scabland flood of the Columbia River Plateau in eastern Washington: Quaternary Research, v. 10, p. 171–178.

_____, and Crandell, D. R., 1978, Potential hazards from future eruptions of Mt. St. Helens volcano, Washington: U.S. Geol. Survey Bull. n. 1383-C, 26 p.

_____, 1970, Tectonic structure of the main part of the basalt of the Columbia River Group, Washington, Oregon and Idaho: U.S. Geol. Survey Misc. Geol. Inv. Map 1-587, scale 1:500,000.

Palmer, L., 1977, Large landslides of the Columbia River Gorge, Oregon and Washington: Geol. Soc. America, Reviews in Engineering Geology, v. 3, p. 69–83.

_____, 1982, Giant river channel forms on Portland, Oregon: Oregon Acad. Sciences (unpub.)

Pardee, J. T., 1910, The glacial Lake Missoula, Montana: Jour. Geology, v. 18, p. 376–386.

_____, 1942, Unusual currents in glacial Lake Missoula, Montana: Geol. Soc. America Bull., v. 53, p. 1569–1600.

Parsons, R. B., and Balster, C. A., 1969, Late Pleistocene stratigraphy, southern Willamette Valley: Northwest Science, v. 43, n. 3, p. 116–129.

_____, Balster, C. A., and Ness, A. O., 1970, Soil development and geomorphic surfaces, Willamette Valley, Oregon: Soil Sci. Soc. Am. Proc., v. 34, p. 485–9.

Patton, P. C., Baker, V. R., and Kochel, R. C., 1978, New evidence for pre-Wisconsin flooding in the channeled scabland of eastern Washington: Geology, v. 6, p. 567–571.

Pierce, J. E., 1982 (pers. comm.)

Playfair, J., 1802, Illustrations of the Huttonian theory: Edinburg.

Read, H. H., 1952, The geologist as historian: (In) Scientific objectives, Butterworth's Scientific Publications, London, vii, p. 58.

Richmond, A. M., 1935, British Columbia's industrial minerals: Canada Min. & Met. Bull. 280, p. 385.

Richmond, G. M., 1965, Glaciation of the Rocky Mountains: (In) Wright, H. E. Jr., Frey, D. D. (Ed.), The Quaternary of the United States, Princeton University Press, p. 217–230.

_____, Fryxell, R., Neff, G. E., and Weis, P., 1965, The Cordilleran ice sheet of the northern Rocky Mountains and related

Quaternary history of the Columbia Plateau: (In) Wright, H. E. Jr., Frey, D. G. (Ed.), The Quaternary of the United States, Princeton University Press, p. 231–242.

Rigby, J. G., 1982, The sedimentology, mineralogy, and depositional environment of a sequence of quaternary catastrophic flood-derived lacustrine turbidites near Spokane, Washington: Univ. of Idaho, M.S. thesis, 132 p.

Rutten, M. G., 1955, Mathematics in geology and the former extension of the Pre-Cambrian (with discussion): Geol. en Mijnbouw, v. 17, p. 192–193.

Santiago, G. T., 1967, Some problems in the physical anthropological study of the peopling of North America: Current Anthropology, v. 8, n. 4, p. 298–299.

Sargent, S. C., 1953, Tree-rafted erratics along the Columbia River: Geol. Soc. Oregon Country News Letter, v. 19, n. 6, p. 63–64.

_____, 1956, Foundation report, The Dalles dam, Corps of Army Engineers, Portland district, Part I, Introduction and spillway.

Scientific American, 1971, Energy and Power: Freeman & Co., 144 p.

Shutler, Richard, Jr., 1983, Early man in the new world, Sage Publications, Beverley Hills, California.

Stauffer, James, 1956, Late Pleistocene flood deposits in the Portland area: Geol. Soc. Oregon Country News Letter, v. 22, n. 3, p. 21–31.

Strong, Emory, 1959, Stone age on the Columbia River: Binfords and Mort, 254 p.

Treasher, R. C., 1938, A Pleistocene damming of the lower Columbia River (abst.): Geol. Soc. Oregon Country Newsletter, v. 4, n. 24, p. 271.

_____, 1942, Geologic history of the Portland (Oregon) area: Ore. Dept. of Geology and Min. Industries, Short Paper 7, 17 p.

Trimble, D. E., 1963, Geology of Portland, Oregon and adjacent areas: U.S. Geol. Survey Bull. 1119, 119 p.

Waitt, R. B. Jr., 1980, About forty last-glacial Lake Missoula jokulhlaups through southern Washington: Jour. Geol., v. 88, p. 653–679.

_____, 1983, Tens of successive, colossal Missoula floods at north and east margins of Channeled Scabland: U.S. Geological Survey, Open-File report 83-671, 29 p.

_____, 1984, Periodic jokulhlaups from Pleistocene glacial Lake Missoula—new evidence from varved sediment in northern Idaho and Washington: Quaternary Research, v. 22, p. 46–58.

_____, 1985, Case for periodic, colossal jökulhlaups from Pleistocene Lake Missoula: Geol. Soc. America Bull. v. 96, n. 10, p. 1271–1286.

Walker, R. G., 1967, Varved lake beds in northern Idaho and northeastern Washington: U.S. Geol. Survey, Prof. Paper 575-B, p. 83–87.

Weis, P. and Newman, W. L., 1971, The channeled scabland of eastern Washington—the geologic story of the Spokane flood: U.S. Geol. Survey pamphlet, 25 p.

Williams, Ira A., 1923, The Columbia River gorge, its geologic history interpreted from the Columbia River highway: Min. Res. of Oregon, Oregon Bur. of Mines and Geol., 130 p.

Winkless, N., and Browning, I., 1980, Climate and the affairs of men: Fraser Publishing Co., 228 p.

Wright, H. E. Jr., and Frey, D. G. (Ed.) 1965, The Quaternary of the United States: Princeton University Press, 922 p.

INDEX AND GLOSSARY

Italicized page numbers are where technical terms are defined. **Boldfaced** page numbers are illustrations, maps, figures, and photographs.